An Introduction to Materials Science

An Introduction to Materials Science

Wenceslao González-Viñas
Héctor L. Mancini

PRINCETON UNIVERSITY PRESS

PRINCETON AND OXFORD

First published in Spain as *Ciencia de los materiales*

copyright © Editorial Ariel 2003

Translation copyright © 2004 by Princeton University Press

Published by Princeton University Press, 41 William Street, Princeton, New Jersey 08540

In the United Kingdom: Princeton University Press, 3 Market Place, Woodstock, Oxfordshire OX20 1SY

Library of Congress Cataloging-in-Publication Data

González-Viñas, Wenceslao, 1969–
　　[Ciencia de los materiales. English]
　　An introduction to materials science / Wenceslao González-Viñas, Héctor L. Mancini.
　　　　p. cm
　　Translation of: Ciencia de los materiales.
　　Includes bibliographical references and index.
　　ISBN 0-691-07097-0 (cloth : alk. paper)
　　1. Materials. I. Mancini, Héctor L., 1945– II. Title.
　TA403.G5313 2004
　620.1′1—dc22　　　　　　　　　2003062200

British Library Cataloging-in-Publication Data is available

www.pupress.princeton.edu

10　9　8　7　6　5　4　3　2　1

To Our Families

Contents

Figures

Tables

Preface

The science of materials is nowadays one of the most significant and active areas of knowledge. Therefore, it is almost impossible to condense into one book an adequate introduction in which the extent and depth of that knowledge are sufficiently reflected. Even so, we make the attempt in this volume.

Contemporary industrial and technological development has been achieved by means of materials constituted of essentially raw substances. The materials have evolved from natural or derived products to current synthetically designed materials whose properties are frequently already defined in advance.

Knowledge from different branches of science such as physics, chemistry, thermodynamics, statistical mechanics, electromagnetism, and quantum mechanics has facilitated the development of the science of materials.

Research on any material requires a synthesis of knowledge, and it is truly amazing how much scientific and engineering work has been done on every material of technological interest over the last century. The progress of recent decades in studying the relation between structures and properties has opened the gates to an astonishing development of new materials and previously unthinkable applications. It is enough to consider the role of a few impurities in electrical conduction in, say, semiconductors to evaluate a material's importance.

Because of a lack of knowledge about the relations between properties and structure, for many centuries technology had to apply trial and error procedures in order to evolve. Today, this knowledge enables us to develop new materials with two industrial advantages: less time and fewer costs. In addition, materials are engineered, that is, their physical, chemical, mechanical, and electrical properties are taken into account. So we can assert that knowledge and technology are strongly associated in the evolution of materials science. Without this close relationship, the costs and time necessary to generate new materials would be enormous. For the reasons expressed above, the great quantity of new materials has helped to increase comfort and has improved communications and health. Summing up, the quality of life has become better.

This book tries to teach undergraduate students—and other people who want to know about fundamental subjects included in the science of materials—enough to connect the properties and structure of materials that today's technology uses. This relation provides the necessary tools to estimate properties by means of calculus techniques available to any science student. Computational methods are essential for this.

Furthermore, at the present time enterprises that produce new materials and research laboratories have developed extensive databases, and thus they can accurately provide the properties of products. For this reason there is no need to repeat here tables of data with no didactic purpose.

Teaching of the science of materials must offer basic principles that permit us, on the one hand, to choose from the sea of data the ones that are correct and suitable for an application, or to calculate them; on the other hand, for new materials designed for a specific aim, teaching must afford the judgment to know which structural elements can yield materials with the properties we desire.

In this book, metallurgical topics such as Fe-C alloys and classical ones like ceramic materials, among others, are missing. We prefer to emphasize materials that are of decisive importance in today's industry, semiconductors, superconductors, optical materials, and the science of surfaces, or frontier topics like fullerenes, quasicrystals, and biocompatible materials. Among these topics are chapters meant to offer an understanding of the fundamentals of the science of materials. Nevertheless, we hope that the absence of classical chapters does not interfere with this book's intelligibility.

Materials of the future will be composites with different organizational levels. Their properties will be measured by selecting molecular components, intermolecular assembly methods, structural order, impurities, and defects. When diversifying and structuring the composition of materials and also working on the structural organization needed for a certain result, humans follow in the steps of nature, where this principle has been achieved to maximum perfection in the constitution of living beings.

We thank M. A. Miranda for drawing so many excellent figures. We also thank Hugo J. Mancini for reading an early version of the book and for helpful comments to improve it. We wish to acknowledge help with the translation of the book given by M. A. Miranda and Pilar Ayucar. And we feel indebted to our families for all their help and understanding, especially Montse and María del Carmen. Finally, during the course of writing this book we very much appreciate the support we have received from the University of Navarre.

Wenceslao González-Viñas
Héctor L. Mancini
May 2, 2003

An Introduction to Materials Science

Chapter One

What Is a Material?

From a practical standpoint, we know that *material* objects are essentially all substances that a human being needs to *build things*. This definition includes solids, but also liquids (e.g., liquid crystals that create LCD displays), and even gases for more specific situations. Really, every raw material used by industry could be included in this classification, but we use the word "material" in a restricted sense: We think about materials whose properties might not be an exact image of those that their elements possess. Thus, we especially concern ourselves with how elements are structured in macroscopic bodies, with how treatments are used during the elaboration of materials, or with the physicochemical aggregation of different elements—all activities that condition the properties of the materials we generate.

The selection, modification, and elaboration of materials to satisfy our needs merge in the foundations of human culture. From the very beginnings of prehistory, humans have manipulated substances so that they would be more useful. To create more useful materials, our forebears wanted to understand and control the composition of materials, and they often succeeded in modifying a material's behavior and properties and in predicting the effects of such manipulations.

This task developed over time, beginning as a handicraft that employed empirical and speculative knowledge. The history of materials science and engineering had already begun in the Stone Age when stones, wood, clay, and leather began to be manipulated. In the Bronze Age, mankind discovered the value of temperature and used it to modify materials by thermal treatments or by adding other substances. Yet, in spite of technological improvements, materials science remained empirical until the end of the nineteenth century. Materials science, as we now understand it, began with the appearance of Mendeléev's periodic table. Since that time, some properties of elements that are related to their position in the periodic table began to be explained scientifically, and these results became incorporated in the annals of science. Since the end of the nineteenth century, the introduction of chemistry and physics, calculus, and modern experimentation have brought the use and profits of materials to a mature status. Currently, thanks to more reliable knowledge of the structure of matter, we can design new materials atom by atom, to achieve the properties we want. At last we have materials that not only satisfy our requirements, but also permit us to create new ones that were hitherto unthinkable.

Thanks to this science, we can even speculate about using new, alternative materials to solve socioeconomic problems by avoiding the decimation of natural resources or trying to reach long-range sustained economic development. Conversely, the solution of unsolved problems improves our theoretical knowledge as well as the scope of materials in science and engineering.

In this context, materials scientists must analyze how the structure and composition of materials relate to their properties, and the effect of the method of preparation of a material. Materials engineers examine the preparation, selection, and application of materials in agreement with known and desired properties. Engineers also incorporate technical and structural analysis and examine key concerns: energetic, economic, ecological, aging.

For materials science and engineering, changes in *physicochemical properties* in response to a stimulus are highly significant. These properties can be classified into groups according to the kind of stimulus: mechanical, thermal, electromagnetic (throughout the spectrum), chemical, and scattering. In brief, mechanical properties, such as deformation and fracture, among others, are responses to applied mechanical forces. Thermal properties, like thermal conductivity and heat capacity, are affected by heat fluxes or temperature changes. Electrical properties such as the dielectric constant or conductivity occur in response to electromagnetic fields. Magnetic properties, like different types of magnetism, are also a response to electromagnetic fields. In a similar sense, optical properties, such as the refractive index or absorption, among others, respond to electromagnetic fields having high frequency. Chemical properties, like the chemical affinity, are responses to the existence of reagents in the environment, and the scattering properties are responses to the impact of particles depending on the material's structure.

In thinking about properties as a response to determined stimuli, we can group materials into families that facilitate a common analysis to determine the origin of the properties. For example, materials can be classified according to their electrical properties; hence, the materials are grouped as good or poor electrical conductors. This brings us to a taxonomy that permits us to see common features between materials in a family, to understand the basis of a property, and to predict the origin of new materials.

In the selective process of materials engineering, the choice of material is limited by the required properties and by the available budget. The requisite properties are imposed by what we wish to make from the material, by environmental conditions, and by the degradation of the material. In this selection we have to take into account that the usage of materials and environmental conditions will provoke their degradation. The degradation of materials determines the required properties in an environment. When environmental conditions can be controlled, material selection is defined by its usage and the budget. That is, the economy plays a key role in materials engineering.

Materials science itself tries to analyze phenomena by the usual activities of contemporary science, and, without relying on economic aspects, to determine how structure, the presence of impurities and defects, production, purification, or mechanical transformation affects material properties.

Materials science can also do the converse: As a group of desirable properties is defined, the material that can display them, although it might not exist in nature, is designed. There are well-known examples of this: stainless steel, powders used in metallurgy, ceramic materials with a controlled coefficient of expansion (which can even be zero), conducting plastics, plastics with a high resistance to friction, such as the one used in some aircraft radomes (a word formed from radar dome), or glasses with a saturable transmission coefficient.

The continuous development of new materials has also prompted the growth of an innovative industrial sector whose products, such as microelectronics or photonics, have greatly transformed the relationship between humans and their environment. Suffice it to say that with the many appliances that are electronically controlled, with the computer industry, with the substitution of copper by optical fibers in telephone conductors, or with satellite communications, we are challenged to make sense of the socioeconomic impact that these changes imply. Countries need to modify their industrial structure so they can survive the modifications that the new materials technology generates.

1.1 CLASSIFICATIONS

The phase of a material—which defines its macroscopic presentation—characterizes the material's properties and depends on external variables like temperature and pressure. This phase can be modified when external parameters are changed. If we want to assert that a sample is of a certain type, we have to specify, apart from the material, the interval of environmental conditions in which its phase is stable. For example, we cannot say that aluminum is a conducting material without specifying the temperature at which it acts like a conductor; this is because at temperatures lower than 1.19 K aluminum reveals a superconducting phase with quite a different phenomenology from the conducting one. Often this is not enough. Metastable states appear because of a material's degradation, hence allowing a sample of a material that is stable under certain conditions to coexist with another sample in another phase under the same conditions. As an example, carbon in normal conditions can naturally coexist in allotropic forms like diamond and graphite. It is not sufficient to indicate the environmental conditions; data about the sample's *history* are also required. In this example the pressures and temperatures applied to the carbon atoms and their duration are required for unambiguous determination of the phase of a sample. Without analyzing such problems, we list below some possible classifications.

The most general materials classification consists of dividing them into *simple materials* and *composites*. Composites are formed of more than one different type of material. After this simple classification, it is common to classify materials according to their different properties; hence, for example, as follows.

- Components:

    - Simple elements: monatomic and polyatomic.

    - Compounds: diatomic, polyatomic, macromolecular (organic and inorganic).

    - Mixtures. These correspond to composites and can be of different chemical compounds or of different phases of the same compound. Blends can be either homogeneous or heterogeneous. The division of mixtures is made with reference to the following scales: atomic, microscopic, mesoscopic, macroscopic (various). For example, it is not enough to assert that a granite sample is heterogeneous; it has to be stated that it is heterogeneous at the 1 mm scale, which is homogeneous at the 1 km scale.

- Type of bond:[1]

    - Ionic (insulators, ceramics, metal-nonmetal). Bond energy 3–8 eV/atom.

    - Ionic-covalent.

    - Covalent (polymers, ceramics, and so on).

    - Metallic (metallic materials). Bond energy from 0.7 (Hg) to 8.8 eV/atom (W).

    - van der Waals: fluctuating induced dipole (H_2, Cl_2, and so on) with bond energy of 0.1 eV/atom, induced dipole-polar molecule (e.g., HCl) with bond energy of 0.1 eV/atom, permanent dipole or hydrogen bond (e.g., H_2O, NH_3) with bond energy of 0.5 eV/atom.

    - Pseudobond or physical bond (*sticky* materials).

---

[1]We will deal with this classification at the beginning of chapter 2.

- Electrical properties:[2]

    - Metallic or conducting, including principally metals and metallic alloys. In conductors the electrical resistance R is low but increases as the temperature rises.

    - Semimetallic. In these, the electrical resistance R is appreciable, but there are 10^{-4} fewer electrons than in the metallic materials. Again, the resistance R grows as the temperature increases.

    - Semiconducting. They have an appreciable electrical resistance R that diminishes if the temperature rises.

    - Insulating or dielectrics. They have a high electrical resistance R.

    - Superconducting. Their electrical resistance is $R \approx 0$.

- Arrangement of components:[3]

    - Monocrystalline.

    - Polycrystalline.

    - Glassy materials, which present short-range order.

    - Quasicrystalline.

    - Semicrystalline.

    - Partial order. For example, the material may have positional order (the mass centers of the components have an ordered disposition) but not orientational order (the components, necessarily anisotropic here, do not have an ordered orientation).

    - Amorphous.

    - Composite (see the classification according to the components).

Because of the existence of many such nonequivalent classifications and of intermediate materials, the classifications above are of limited value. We assume, usually, the following classification, but for our convenience and for didactic reasons sometimes we will use either one or another of the preceding classifications.

1.2 FUNDAMENTAL PROPERTIES OF DIFFERENT KINDS OF MATERIALS

Although the properties of the materials in each category can sometimes vary, properties broadly accepted as defining the categories are the following:

- Metallic materials:

    - They are built up of metallic elements or of compounds of metallic elements.

    - They have many unlocalized electrons in the so-called conduction band.

    - They are good thermal and electrical conductors. They are opaque to visible light.

    - They are usually strong and plastic.

---

[2]We will use this classification in sections 2.5 and 6.5 and in chapters 4 and 7.
[3]We will use this classification in sections 12.2 and 10.3 and in chapters 2, 3, and 9.

- Ceramic materials:

  - They are chemical compounds of the type metal + nonmetal.
  - They are generally good electrical and thermal insulators.
  - They are stronger than metallic and polymeric materials at high temperatures and in chemically aggressive environments.
  - They are hard and brittle.

- Polymeric materials:

  - They are compounds, generally organic, in the form of long chains.
  - They have low density.
  - They are flexible or elastic or both.

- Semiconductor materials:

  - They have properties intermediate between conductors and insulators.
  - They have properties that are extremely sensitive to impurities and to temperature.

- Composites:

  - They are composed of more than one type of material.
  - They are designed to obtain better properties or combinations of properties. For example, glass fiber is as resistant as glass filament and as flexible as the polymer that forms it. Another example is adobe. Adobe, a mixture of clay and straw (up to 30%), has been employed in making bricks and in primitive buildings. Composites also serve as new materials in the aerospace industry. Together, these are the two technological ends of composites.
  - They are designed by taking into account typical targets of materials engineering. As an example, the aims of the aerospace industry are first to reduce operating costs (i.e., to reduce the mass of structures to increase the payload); and second to raise the fatigue limit, stiffness, toughness, and thermal shock resistance. It is almost impossible to attain both aims with simple materials, so one must take into account composites like graphite-epoxy, among others.

Chapter Two

Crystalline Solids

Under normal environmental conditions, we classify materials by their solid, liquid, or gaseous states of equilibrium, depending on how they aggregate. In this book we concentrate mainly on the *solid state*—the state of matter that has a determined volume and shape. In the *liquid state*, matter has a determined volume but not a shape, and in the *gas state*, matter lacks a defined shape and fills the entire volume of available space. It is also common to add another state to this classification, the *plasma state*.[1] Here, matter appears to be completely ionized. Nevertheless, for materials scientists, this state is of no practical interest, other than in some preparation methods.

Coexisting with these four states of matter are many more *phases* that display varied macroscopic forms of materials (e.g., crystalline phases, gas phases, amorphous phases, anisotropic fluid phases [cholesteric, nematic, etc.], and so on). Some of these phases, such as the glassy one, belong to a border zone between states of matter, a zone in which the exact border is ambiguous. When these phases diverge from our definitions, we must specify parameters, such as the observation time scale, to classify them into one state of matter.

Although matter is found in any of the three states of aggregation, only with solid materials does spatial order gain real prominence, as do the atoms, ions, and molecules contained by solids. Their spatial structure, the kind of order or disorder in which the chemical elements are arranged, also helps to define their properties. Yet now we can design materials that exploit the properties of different types of chemical components, including arrangements or matrices of materials that define the spatial structure. Such chemical elements can be introduced either as key ingredients or as impurities. They enable one to come up with materials whose properties were previously chosen. Probably the most typical example in the twentieth century is the semiconductor, whose dominant role is unquestioned. Its discovery has enhanced the nature of communications, the automatic production of goods by means of robots, the matériel of the military, and the intellectual support that occurs with computers and calculators. Technological improvements have opened the doors to smaller and more powerful semiconductor devices, which in turn have transformed our homes into places replete with microwave ovens, sound systems, televisions, and video recorders.

If the chemical structure of atoms and molecules is essential to the properties of solids, it is even more important for liquids and gases, where the chemical structure is the sole determinant of all other properties. Hence the science of materials focuses more on the relation between the spatial structures of materials and their properties in the solid state than in liquids and gases.

In a solid, atoms or molecules are close packed and are maintained in positions fixed by means of electromagnetic forces that have the same order of magnitude as those in molecular bonds. In solids, atoms or molecules are not like isolated entities. On the contrary, their

---

[1]Also, there is the Bose-Einstein condensate state.

properties are modified by their proximity to other atoms or molecules, which modify the energy levels of their outer electrons. Solids whose structures have spatial regularity or periodicity are known as crystalline; solids that have no order are called amorphous (the experimental determination of these structures is briefly treated in section 2.6). There are also intermediate types of materials, as shown in section 1.1. A complete theory of these materials must correlate macroscopic properties like elasticity and hardness, electrical and thermal conductivity, and optical reflectivity with their spatial structures.

Some general properties are due to the kind of bond between the solid constituents. This is the initial criterion for the classification that we use.

1. *Ionic materials* (ionic crystals). A regular distribution of positive and negative ions results when some valence electrons are transferred from one component to another (figure 2.1), which is the case for NaCl, KCl, and CsCl, among other materials. The atoms are distributed in a stable way because of the very strong electric interactions. For example, the distance between Na^+ and Cl^- in NaCl is 2.81×10^{-10} m. Ionic materials are poor conductors of heat and electricity, hard, brittle, and have a high melting point. The atoms can absorb energy in the far infrared (< 1 eV), creating, for example, a vibrating mode in the crystal lattice.

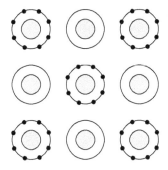

Figure 2.1 Scheme of an ionic solid.

2. *Covalent materials.* As we see later (figure 2.8), there is a continuous gradation between ionic and covalent bonds. Nevertheless, a pure covalent bond joins atoms directionally, as with covalent molecules (figure 2.2). A typical case is diamond, an allotropic form of C, in which the four bonding electrons of each atom of C are localized according to tetrahedral sp^3 wave functions that bind each atom of C to the other four; the result is the

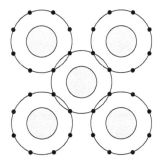

Figure 2.2 Scheme of a covalent solid.

crystalline structure of diamond. The separation between two atoms of C is 1.54×10^{-10} m. In a way, covalent solids are similar to three-dimensional macromolecules. They are very hard and poor conductors, and the necessary energy to excite vibratory modes in the lattice is high due to the bond's rigidity and directionality. The energies of excited electronic states are much greater than are thermal energies (of the order of $k_B T$, approximately 20 meV at room temperature) or the energy of visible light (1.8–3.1 eV). Consequently, these materials are usually transparent.

3. *Metallic materials.* These materials are composed of elements that have low ionization energy and few electrons weakly bound in their incomplete outer shells. The electrons are easily released when a solid is formed; this provides an ionic lattice inside a gas of electrons that has relatively high mobility (figure 2.3). These are good conductors and opaque owing to the characteristics imposed by the electrons in the gas, which are independent of the underlying ionic lattice.

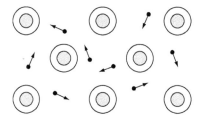

Figure 2.3 Scheme of a metallic solid.

4. *van der Waals materials.* These materials include those with a *hydrogen bond* such as H_2O and HF. They possess great polarity (figure 2.4). This behavior arises because hydrogen has only one electron, which does not totally screen the proton's charge, and it interacts with other molecules to form a solid. Also, van der Waals materials include *molecular* ones generated by van der Waals interactions (figure 2.5) such as Cl_2 and CO_2. These bonds are the same as hydrogen bonds but weaker. All of these solids have a low melting point and, like the bond they are made of, are soft and weak.

Figure 2.4 Scheme of a hydrogen bond solid.

5. *Materials with various types of bonds.* This is the case for graphite, an allotropic form of carbon. These materials have partially covalent bonds owing to their hybrid sp^2 wave functions, which are directional and localized, and partially van der Waals bonds due to their p_z wave functions, which are weak and delocalized. Thus, they have special properties, as we explore below.

In this chapter we present the basics of crystalline solids in order to understand their properties and uses.

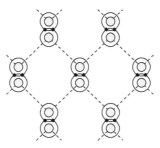

Figure 2.5 Scheme of a van der Waals solid.

2.1 ORDER, TRANSLATIONAL INVARIANCE, AND ANISOTROPY

The concept of *order* is fundamental in the science of materials. It indicates, as a first approximation, a good arrangement of things. "Goodness" indicates the ease with which one can determine the nature of a material's arrangement. That is, if we know the arrangement (position, state, etc.) of an element, we can say that the elements have a readily determined number of neighbors or contiguous elements; these, too, have arrangements (position, state, etc.) that can be determined. In the science of materials, elements (whether they are ordered or not) can be atoms, ions, molecules, etc.; we will call them components. From the order's *range* we know the characteristic distance for which the order holds true. If the range is much greater than the interatomic distances, the order is long range. Examples are monocrystals of macroscopic size and polycrystalline samples. Also, there is short-range order for instances where we can measure the arrangement of the components nearest to a determined one; this is so for glassy materials (common glass, for example). Another type of material is amorphous, which presents itself in totally disordered forms. Here we can characterize the components' disposition only with statistical methods.

In the science of materials the mathematical meaning of order is usually related to the concept of *translational invariance*. Nevertheless, it does not always occur, as we find in chapter 9. Let us define a translation vector $\vec{d}$ as a linear geometric operation. The translation is assigned to each space vector, which is the same vector plus the translational vector: that is, $\forall \vec{x}, \quad \vec{x} \mapsto \vec{x} + \vec{d}$. According to this definition, a system does not vary during continuous translations of the unit vector $\hat{d}$ if $\forall \epsilon \in \Re$ and $\forall \vec{x}$ the system behaves physically and chemically in the same way in $\vec{x}$ as in $\vec{x} + \epsilon \hat{d}$. A system is thus invariant during discrete translations of the vector $\vec{d}$ if $\forall \vec{x}$ the system behaves physically and chemically in the same way in $\vec{x}$ as in $\vec{x} + \vec{d}$. This condition is satisfied with crystalline materials, where we have a finite set of components that are periodically repeated according to translational vectors in different directions in space. If a system does not vary during continuous translations in every spatial direction, it is *homogeneous*. Granted that the measurement or observation equipment determines the length scale of the spatial resolution, the concept of homogeneity is closely linked to that scale, at least from a functional point of view.

Isotropy is also an important concept, one related to the preceding concepts. A system is isotropic if its physicochemical properties are independent of the direction in which they are observed. If the system is not isotropic it is anisotropic. The following example makes this easier to understand. Let us consider gelatin. If we try to cut it with a knife we will make the same stress independently of the direction in which we try to cut it. In an anisotropic system, however—such as concrete reinforced with steel bars—it is easier to cut or break the material in directions parallel to the bars rather than in perpendicular ones. Thus, in general, ordered materials may display anisotropy. The condition for isotropy is reflected mathematically in the sense that the system's properties depend not on the vectors but only on their moduli. In general, *anisotropy* may be attributed to physicochemical properties if they depend on the sample's direction. All this allows us to specify the general anisotropy that a sample presents; that is, the sample displays anisotropy for one property but not for others.

Anisotropy can be due to various causes:

- Anisotropy of components. This occurs when atoms, ions, or molecules in a material sample are anisotropic. This elemental anisotropy can induce sample anisotropy, as with a liquid crystal in the nematic phase (see section 12.2), or it may not, as with amorphous silica.

- Anisotropy of structure. This is also called crystalline anisotropy and occurs when the disposition of atoms, ions, or molecules in the sample depends on the direction. This anisotropy can also induce sample anisotropy, as in the monocrystalline sample of a material, or it may not, as in an amorphous sample of the same material.

- Shape anisotropy. This occurs in samples that have planar defects in the broadest sense of the term (see section 3.4). In particular, shape anisotropies can appear in a sample because it has an appreciably finite size and therefore external surfaces, or because the sample is polycrystalline and therefore has grain limits. This anisotropy can also induce sample anisotropy, as with a magnetized ferromagnetic sample, or it may not, as we find in a large enough sample of polymethyl methacrylate.

- Induced anisotropy. This anisotropy is due to the presence of external agents that modify, transiently or permanently, the components, the structure, or the planar defects of the sample in such a way that some properties are a function of the material's direction. This may induce anisotropy in a sample. For example, a moderate magnetic field applied to a paramagnetic material creates anisotropy in the sample's magnetization, or it may not induce it; for example, a weak electric field applied to a metallic material does not produce anisotropy for thermal properties.

- Space anisotropy. The principle of *space isotropy* is considered here. This principle enables one to define equivalent directions in a crystallographic lattice because from a physicochemical standpoint the directions are regarded as indistinguishable.

2.2 BRAVAIS LATTICE, WIGNER-SEITZ CELL, RECIPROCAL LATTICE, AND FIRST BRILLOUIN ZONE

A *Bravais lattice* can be defined as a discrete arrangement of points in space—called lattice *sites*—in such a way that this arrangement is always the same independently of the site from which we see it. That is, the lattice presents discrete translational invariance in the three independent spatial dimensions. There is also an alternative definition: A d-dimensional Bravais lattice is the set of points, the so-called lattice sites, defined by

$$\vec{R} = \sum_{i=1}^{d} n_i \vec{a}_i,$$

where the vectors $\vec{a}_i$ are linearly independent and are known as primitive vectors of the lattice, and n_i are any integer numbers. Note that the set of primitive vectors of a lattice is not unique. It is possible to have complementary interpretations of these definitions: One is that the Bravais lattice represents the set of lattice sites directly; another is that the Bravais lattice represents the set of vectors or translations that leave the lattice invariant. The Bravais lattice is also called the *direct lattice*.

Some examples follow here.

- The body-centered cubic lattice (bcc) (figure 2.6). If a is the length of the cube edge, called the *lattice constant*, a set of primitive vectors is $\vec{a}_1 = a\hat{i}, \vec{a}_2 = a\hat{j}, \vec{a}_3 = (a/2)(\hat{i} + \hat{j} + \hat{k})$ (problem 2.1). Another possible set of primitive vectors is $\vec{a}_1 = (a/2)(-\hat{i}+\hat{j}+\hat{k}), \vec{a}_2 = (a/2)(\hat{i} - \hat{j}+\hat{k}), \vec{a}_3 = (a/2)(\hat{i}+\hat{j}-\hat{k})$ (problem 2.2).

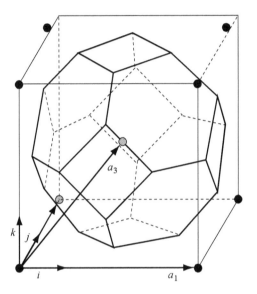

Figure 2.6 A primitive vector set and the Wigner-Seitz cell of a bcc lattice.

- Face-centered cubic lattice (fcc) (figure 2.7). This is also a Bravais lattice. A possible set of primitive vectors is $\vec{a}_1 = (a/2)(\hat{j} + \hat{k}), \vec{a}_2 = (a/2)(\hat{i} + \hat{k}), \vec{a}_3 = (a/2)(\hat{i} + \hat{j})$ (problem 2.3). Another set of primitive vectors is $\vec{a}_1 = (a/2)(\hat{i} - \hat{k}), \vec{a}_2 = (a/2)(\hat{j} - \hat{k}), \vec{a}_3 = a\hat{k}$ (problem 2.4).

We can associate with each lattice site its nearest neighbors; these are the lattice sites closest to the first one. In this context, the *coordination number* is the number of nearest neighbors in a lattice site.

The concept of the *unit cell* is also important. It represents the space volume, so if we apply lattice translations to it we can reproduce the whole lattice without overlaps or empty spaces. A *primitive unit cell* (figure 2.7) is a unit cell in the sense that to build the whole lattice we must use *every* lattice translation.

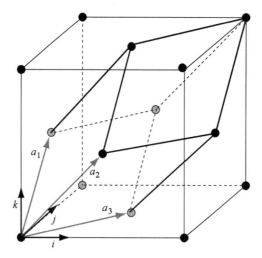

Figure 2.7 A primitive vector set and a primitive unit cell of an fcc lattice.

With these two concepts we can show that

- The paralleliped formed by primitive vectors is a primitive unit cell,

- The primitive unit cell is not unique, and

- Every primitive unit cell has the same volume.

The *Wigner-Seitz primitive cell* (figure 2.6) is the minimum volume formed by space points closer to the lattice site considered than to any other one. The crystal basis is the physical unit located at every lattice site. Thus a bcc lattice can be regarded as a simple cubic lattice if we choose the following basis: $\{0, (a/2)(\hat{i} + \hat{j} + \hat{k})\}$.

Not every crystalline lattice is a Bravais lattice. In fact, only 14 three-dimensional Bravais lattices occur in a total of 230 crystalline lattices. The concept of the basis also enables us to describe crystalline lattices that do not correspond to Bravais lattices. Thus, for example, the zinc blende structure lattice (SZn) corresponds to two fcc lattices separated by one-quarter of their diagonal length; thus, the structural lattice is a fcc lattice whose basis is $\{0, (a/4)(\hat{i} + \hat{j} + \hat{k})\}$. For SZn the first elements of the basis are Zn atoms and the second are S atoms.

We are now ready to define some important concepts that elucidate the properties of crystalline solids. The *reciprocal lattice* is the set of vectors $\{\vec{k}\}$, from the so-called reciprocal space, being $e^{i\vec{k}\cdot\vec{r}}$ a periodic function according to the Bravais lattice. Thus, if $\{\vec{R}\}$ represents the vectors of the Bravais lattice, the reciprocal lattice (which is of the Bravais type) will be the set of vectors $\{\vec{k}\}$. These verify that for every vector of the Bravais lattice (direct lattice) $\vec{R}$ an integer number n exists such that $\vec{k} \cdot \vec{R} = 2\pi n$. The reciprocal space can be represented mathematically as a Fourier space; so a site of the reciprocal lattice represents a periodicity of the direct lattice. Then the reciprocal lattice is definable as the wave vectors $\{\vec{k}\}$ of a plane wave with the periodicity of the direct lattice. This alternative definition permits two interpretations of the reciprocal lattice: first, a fundamentally geometrical one where sites in the reciprocal lattice represent parallel crystallographic planes and hence their periodicity; second, a fundamentally physicochemical one, where sites represent wave vectors of periodic plane waves in the lattice.

Reciprocal lattices exhibit the following properties:

- The reciprocal lattice of a reciprocal lattice is itself (problem 2.6).

- The bcc and fcc lattices are mutually reciprocal (problem 2.7). A simple cubic lattice is reciprocal with itself, except for a scale factor.

As the Bravais lattice admits more than one set of primitive vectors, we can conveniently assign to one of these sets of the direct lattice one and only one set of primitive vectors in the reciprocal lattice in order to be consistent with the reciprocity. Then, if the primitive vectors of the direct lattice are $\vec{a}_1$, $\vec{a}_2$, and $\vec{a}_3$, the vectors chosen as primitive vectors of the reciprocal lattice *associated* with the first ones are

$$\vec{b}_1 = 2\pi \frac{\vec{a}_2 \times \vec{a}_3}{\vec{a}_1 \cdot \vec{a}_2 \times \vec{a}_3}, \quad \vec{b}_2 = 2\pi \frac{\vec{a}_3 \times \vec{a}_1}{\vec{a}_1 \cdot \vec{a}_2 \times \vec{a}_3}, \quad \vec{b}_3 = 2\pi \frac{\vec{a}_1 \times \vec{a}_2}{\vec{a}_1 \cdot \vec{a}_2 \times \vec{a}_3}.$$

In this case,[2]

$$\vec{a}_i \cdot \vec{b}_j = 2\pi \delta_{ij}.$$

The *first Brillouin zone* is the Wigner-Seitz primitive cell of the reciprocal lattice. Its volume is $(2\pi)^3/V$, where V is the volume of the primitive unit cell of the direct lattice (problem 2.8).

2.3 CRYSTALLINE STRUCTURES: TYPES OF CRYSTALS AND THEIR PROPERTIES

A material presents some physicochemical properties according to the type of bond; this factor by itself sets a determined order. Therefore, the crystalline structures are strongly related to the solids' properties.

As we noted, materials can be classified according to their range of order; that is, monocrystalline, polycrystalline, glassy, and amorphous materials, among more complex ones, each having a smaller range than the preceding one.

Monocrystalline materials are characterized as perfect crystals whose size is that of the sample. *Polycrystalline* materials have a defined order on a scale much greater than the interatomic distances, without being monocrystals, *glassy* materials have a defined short-range order (a few interatomic distances), and *amorphous* materials are completely disordered.

Monocrystalline materials are found in nature, yet can also be artificially produced. Their creation is careful and slow. Polycrystalline materials are the most common in nature.

When we have a melt and then cool it down, what generally happens is that any local fluctuation of temperature, pressure, or concentration starts the material solidification at the place where fluctuation appeared, the *solidification nucleus*. These fluctuations are statistically produced at the same time at many points, so the material will start to solidify at many nuclei at the same time. What follows is the growth of the crystalline material around these *seeds* (a seed is a place where solidification is favored because it is a solidification nucleus or because the solidification has already begun at that point). But the crystalline material grown around every seed is randomly oriented due to the space and the melted

---

[2] δ_{ij} is the Kronecker delta, which is equal to 1 if its indexes are equal and 0 if they are different.

material isotropies; therefore, the different crystalline grains will not match each other well and they will form borders called *grain boundaries*.

With glassy or amorphous materials the cooling processes are too fast to allow atoms to place themselves at equilibrium positions or lattice sites. These materials are often in metastable states that may transform—sometimes with characteristic times of thousands of years—into their crystalline versions. For example, to achieve an amorphous metallic alloy, a jet of the melted material is aimed at a refrigerated rotary disk at high speeds, so the material solidifies into small pieces with short-range order.

As we will discover in section 2.6, crystalline order can be studied through X-ray diffraction.

A first step in studying crystalline structures is to consider the optimum packing of spheres (atomic or ionic isotropy). For this packing the interatomic distances and the ionic radii are important magnitudes; they are particularly significant in the selection of the crystalline structure, mostly for materials composed of atomic species having different radii.

Suppose that we have a material composed of only one atomic species. Here, the most densely packed structures are the fcc and the cph (close-packed hexagonal) (table 2.1). These packed structures (above all the structures that have a larger atomic packing factor) are characteristic of materials with van der Waals or metallic bonds, when the atoms have closed shells and thus are regarded as hard spheres.

Table 2.1 Coordination number (CN) and atomic packing factor (APF) ($V_{atom}/V_{unit\ cell}$) of various crystal lattices, where R is the atomic radius and LC the corresponding lattice constant

Lattice	CN	LC	APF	Examples
fcc	12	$2R\sqrt{2}$	0.74	Zn, Cd, Ti, Mg, Co
cph	12		0.74	Zn, Cd, Ti, Mg, Co
bcc	8	$4R/\sqrt{3}$	0.68	Cr, Fe, W
sc	6	$2R$	0.52	Po

Ionic crystals are composed of more than one species and follow approximately these rules:

- The nondirectional ionic bond favors the close packing most consistent with the geometry;

- If the anion and the cation cannot *touch*, the structure is unstable;

- The ratio between the ions' radii determines, approximately, the crystalline structure (table 2.2).

Table 2.2 Relation between the crystal structure and the atomic radii

Lattice	CN	Radius ratio	Examples
bcc (2sc)	8	$r^+/r^- \geq 0.732$	CsCl
2fcc	6	$r^+/r^- \geq 0.414$	NaCl
2fcc 1/4 shifted	4		ZnS (mixed), BeO

Diamond has the ZnS structure; however, it has a covalent bond. The structure of covalent bonds is derived from the shape of the electronic wave functions, which increases the possible number of structures. There is continuous gradation from a purely covalent to a purely ionic crystal, which considerably alters the shape of the electronic wave function (figure 2.8) and therefore modifies the crystalline structure.

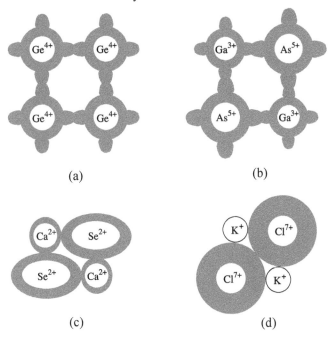

Figure 2.8 Schematic representation of the continuity of the transition from (a) a covalent bond (Ge) to (d) an ionic bond (KCl), going through (GaAs(b)) and (CaSe)(c).

Graphite has a mixed bond (covalent within planes and van der Waals between planes) (figure 2.9) responsible for its metallic character, which is strongly anisotropic, and for its low hardness, which makes it useful for lubricating.

Elements from the VB group of the periodic table (Bi, Sb) have metallic behavior because their intersheet bonds are predominantly metallic. Covalence lowers the coordination

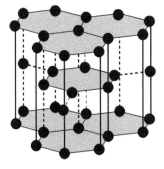

Figure 2.9 Crystalline structure of graphite.

number; hence, in the liquid state, covalence is lost and there is close packing, which means that such materials contract when they melt. A material presents *polymorphism* if it has different crystalline structures. *Allotropy* is the polymorphism for materials made of only one element.

To classify crystalline structures, we divide them into seven *crystalline systems*, each having a different number of symmetries. The crystalline systems (figure 2.10) are cubic, tetragonal, orthorhombic, monoclinic, triclinic, hexagonal, and trigonal.

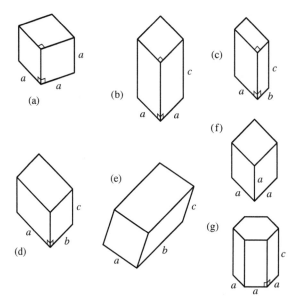

Figure 2.10 Crystalline systems. (a) Cubic, (b) tetragonal, (c) orthorhombic, (d) monoclinic, (e) triclinic, (f) trigonal, and (g) hexagonal.

The number of crystalline systems is equal to the number of point groups of transformations in the lattice (transformations that always leave invariant one point [or more] of the lattice) with a basis of spherical symmetry. If the basis does not have spherical symmetry, we obtain the 32 crystallographic point groups. The number of Bravais lattices is the number of three-dimensional (3D) space groups with a basis of spherical symmetry that convert into the 230 space groups which represent all possible crystalline structures. The number of Bravais lattices for each crystalline system is 3, 2, 4, 2, 1, 1, and 1, respectively, and they present a hierarchical structure where each symmetry group of a system contains all its descendents (figure 2.11).

An ordered triad of numbers $[nms]$ specifies a crystalline direction, and this triad represents the Bravais lattice vector with components n, m, and s in the basis determined by the primitive vectors. A bar is placed above the component to express a negative component. The *Miller indices* specify the crystallographic planes; these indices are each of the components of an ordered triad of numbers (nms) and represent the components of the reciprocal lattice vector normal to the plane in the basis of the reciprocal lattice's primitive vectors. An ordered triad $\{nms\}$ represents the set of Miller indices that correspond to equivalent planes due to crystal symmetries other than translational, much in the same way that a triad $\langle nms \rangle$ represents the set of equivalent crystallographic directions.

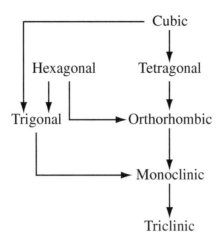

Figure 2.11 Hierarchical structure of the seven crystalline systems. The arrows indicate increased specialization in the symmetry groups.

2.4 SCHRÖDINGER EQUATION, PERIODIC POTENTIALS AND BLOCH'S THEOREM, AND ENERGY BANDS

The classical description of a crystalline solid is based on regarding it as a continuous medium. That is, a length range exists where the lengths are much smaller than the system's characteristic macroscopic lengths and much larger than the length scale where the variations of the number of particles per unit of volume are very large. With this approximation, solids can be studied by means of the balance equations (which tell us which external or internal mechanisms impede the conservation of quantities such as linear momentum, angular momentum, energy, etc.; the mechanism may be an external force, or an external pair of forces, or an energy dissipation, and so on). These balance equations must be complemented with constitutive relations that describe materials. For example, they tell us the response to a certain stimulus (stress deformation). These constitutive relations can be obtained from general symmetry considerations. What is left undetermined is a set of parameters that can be experimentally measured or deduced by means of microscopic theories. These microscopic theories must take into account quantum effects, because the electronic energy levels of isolated atoms are quantized. In contrast, some properties of crystalline solids are not reproducible by means of macroscopic theory—for example the properties of X-ray scattering by crystalline solids—and rely on the fact that matter and energy are quantized.

The nonrelativistic behavior of a particle without *spin* in a field with a defined potential energy $V(\vec{r})$, is described by the wave function ψ, which is a solution of the time-dependent *Schrödinger equation*:

$$\left[-\frac{\hbar^2}{2m}\nabla^2 + V(\vec{r})\mathbf{1}\right]\psi = \imath\hbar\frac{\partial\psi}{\partial t}.$$

If we consider a system in the stationary state, or time independent, the Schrödinger equation converts into the known eigenvalue equation

$$\mathbf{H}[\psi] = E\psi,$$

where $\mathbf{H}$ has a kinetic part ($\mathbf{T} = \vec{\mathbf{P}}^2/2m$, with $\vec{\mathbf{P}} = -\imath\hbar\nabla$), and a potential one $[V(\vec{r})\mathbf{1}]$, and E is the energy corresponding to the eigenvector state ψ of the Hamiltonian operator

H. ψ physically represents a probability amplitude. Then the probability density will be given by $|\psi|^2$, which is normalized to unity.

If our system is composed of more than one particle (typically, N), then the system will be described by a kinetic part

$$\mathbf{T} = \sum_{i=1}^{N} \frac{\vec{\mathbf{P}}_i^2}{2m_i}$$

and a potential one

$$V(\vec{r}_1, \ldots, \vec{r}_N)\mathbf{1}.$$

Now, the square of the modulus of the wave function will be the joint probability density for N particles. For particles with spin, the wave function multiplied by the eigenvector of the spin operator must be symmetrized or antisymmetrized depending on whether we have bosons or fermions, which are particles with integer or semiodd spin, respectively, and also have the characteristic of obeying or not, respectively, Pauli's exclusion principle.

Let us suppose, then, that we have a crystal at 0 K temperature. This implies that the nuclei positions are fixed. It is supposed that the nonvalence (inner) electrons form a fixed ion in the lattice. As an artifice, an approximate or effective potential can be built, and this behaves as if we had taken into account the effect of all the electrons except one. This potential remains periodic. In consequence, we have approximated our intractable problem of N multielectronic atoms with the simpler problem of an electron in a periodic field.

What is the most important characteristic in a crystalline lattice? If we have N identical isolated atoms, for example of the IA group, with one valence electron, we suppose that this valence electron is in the ground state. If the atoms get closer what could happen? As electrons are fermions, Pauli's principle of exclusion acts. Thus, if electrons are not isolated, they cannot be in the same initial state; they occupy levels slightly deviating from the original one. The closer they are (and hence the more Pauli's or interchange interaction occurs among them), the more they deviate. If we have infinite atoms we obtain a continuous band of energy levels instead of discrete levels. So the origin of the *energy band* is a set of atoms (whether they are ordered or not) at the thermodynamic limit with electronic wave functions overlapping spatially. The wave functions were initially in the same atomic state (figure 2.12).

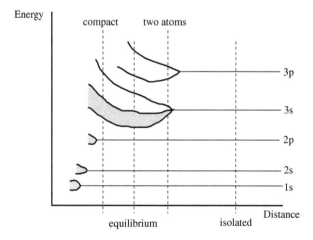

Figure 2.12 Energy bands in sodium depending on the interatomic distance.

For our approximation, we maintain the characteristic formation of bands and simplify it by imposing a crystalline lattice.

The *Bloch theorem* indicates the form of the solution of the Schrödinger equation in periodic potentials. It states that *the solutions of the time-independent Schrödinger equation for periodic potentials with translations corresponding to a Bravais lattice are the product of a periodic function with the lattice periodicity and $e^{i\vec{k}\cdot\vec{r}}$.*

Thus, for each $\vec{k}$ we have states (two, because we have the possibility of two different spin states) in the energy band with their corresponding energies $E(\vec{k})$ (figures 2.13 and 2.14); a particle in one of these states has an effective mass that is inversely proportional to the energy band curvature. This might be different from the electron mass. Furthermore, the mass is anisotropic and depends on the direction considered. In addition, the particle acceleration can have a direction different from the direction of the applied force. Also, $E(\vec{k})$ is periodic and has the periodicity of the reciprocal lattice. When the particle's effective mass is negative it is called a *hole*. When it is positive, it is called an *electron*.

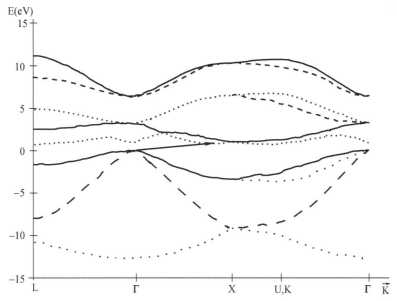

Figure 2.13 Energy bands in germanium for different directions of the reciprocal space. The forbidden zone is between the thick solid curve and the nearest dotted line above. The band gap is indirect, and the minimum-energy transition is indicated.

Therefore, crystalline solids have a *band structure* that consists of a quasicontinuum of states that are occupied by the valence electrons of each of the atoms that form the solid. Between these energy bands the regions with no states are called *forbidden zones*.

The available valence electrons in crystalline solids are distributed in the allowed states of the energy bands so there should be a maximum of one electron per allowed state due to Pauli's exclusion principle. Since each electron has two possible spin states, there might be two electrons for each wave vector. In a first approximation, if we think that the energies of the wave vector states $\vec{k}$ increase monotonically with the modulus of the wave vector, then in the ground state (at $T = 0$ K) the filled states will be limited by an ellipsoidal surface in

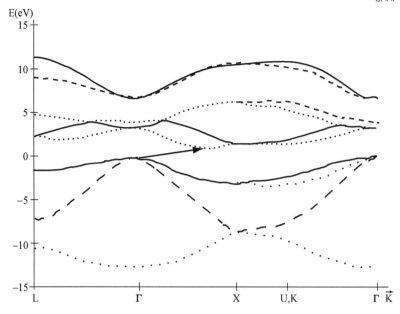

Figure 2.14 Energy bands in silicon for different directions of the reciprocal space. The forbidden zone is between the thick solid curve and the nearest dotted line above. The band gap is indirect and the minimum-energy transition is indicated.

the reciprocal space called the *Fermi surface*. The Fermi wave number is the one associated with this surface.

If the crystal field effect is null (a reasonable approximation for electrons in the conduction band far from its edges), then the electrons in a solid form an interacting gas called a Fermi gas; in addition, the expected isotropy makes the surface spherical. If we consider that the electrons do not interact between themselves, we have a free-electron Fermi gas. In this instance, the electrons have a homogeneous distribution among wave vectors that corresponds to a density of states in the reciprocal space of the volume of the unit cell divided by $(2\pi)^3$. Hence, the number of electrons that get into the Fermi sphere is the Fermi sphere volume multiplied by the state's density and by 2 (due to the spin states). This number is equal to the number of valence electrons in the solid.

2.5 CONDUCTORS, INSULATORS, AND SEMICONDUCTORS

Because these three kinds of materials are characterized by their electromagnetic properties, we can distinguish them when we observe their band structure behavior when an external field is applied.

In this section, energy bands are represented in direct space. This simple representation disregards significant details of the energy bands but confers on them the most simplified interpretation because the direct space is more intuitively understood than is the reciprocal one. To represent energy bands in direct space requires their projection in the reciprocal space (functions of $\Re^3 \mapsto \Re$), where at each point they yield multivalued correspondences to $\Re$ that are the characteristic *bands* (figure 2.15).

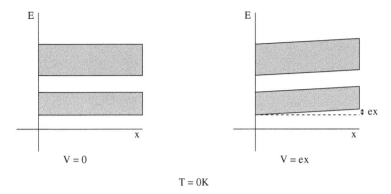

Figure 2.15 Field effect on energy bands.

We shall accept the hypothesis that the external potential varies slowly (on the lattice's characteristic length scale). In this instance it is useful to assume that the external potential modifies only the energy band origin at each point (figure 2.15).

To begin with we examine the case $T = 0$ K, where the system is in the ground state (lowest allowed energy).

For an atom with metallic behavior, the electron shells are semifilled. Therefore, a metallic solid has a semifilled band (figure 2.16) called the conduction band. Here the global electric current is 0 since $E(\vec{k}) = E(-\vec{k})$, and the electronic levels are filled with increasing energy. If we apply a constant electric field with a battery (figure 2.16) we achieve an electric current.

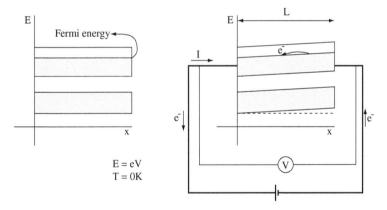

Figure 2.16 Energy bands in a metallic material.

In a semimetal, the valence band maximum is slightly higher than the conduction band minimum. It follows, then, that in the ground state the valence band is not completely filled and the absent electrons are in the conduction band; as a consequence it is possible to have conduction in an electric field (figure 2.17). Because of the small number of these electrons, the electrical conductivity in a semimetal is much less than the electrical conductivity in a conductor.

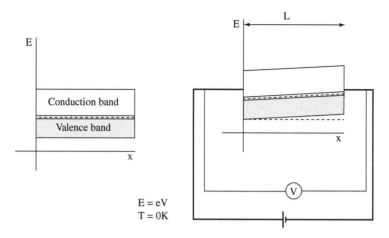

Figure 2.17 Energy bands in a semimetal.

In an insulator the valence band is filled and the conduction band is empty. The two are separated by a forbidden zone whose magnitude is equal to the band gap. If we apply an electric field at temperature 0 K, there is in principle no electronic conduction (figure 2.18).

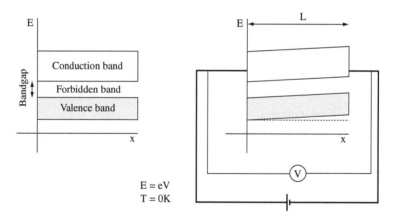

Figure 2.18 Energy bands in an insulator.

Thus, electrons cannot conduct in an insulator except in the following conditions.

1. The electric field should be large and have a range of energies where the conduction and valence bands are in different regions; hence the tunneling effect will exist between the two regions. Furthermore, conduction, albeit small, should be produced (figure 2.19). In these conditions conduction should be present because of an indirect transition; that is, an electron in the valence band elastically acquires linear momentum upon traveling to the conduction band by means of another particle or of an elemental excitation such as a lattice vibration (phonon) that acquires more linear momentum merely to conserve the total momentum.

2. The temperature should be high enough that thermal vibration promotes electrons from the valence band (leaving holes in it) to the conduction band. This, in turn, creates an electric current when an electric field is applied. Increasing temperature also increases

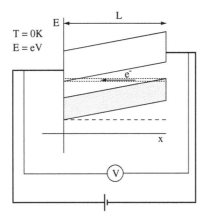

Figure 2.19 Tunneling effect in energy bands in an insulator.

the probability of conduction by the tunneling effect. We can therefore infer that if the temperature rises (in insulators) conductivity increases, because there will be more electrons with enough energy to enter the conduction band (figure 2.20) directly or with the help of other particles or elemental excitations.

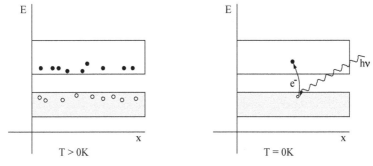

Figure 2.20 Electronic promotion in insulators.

3. It is also possible that electronic promotion can be produced by energy transmission between a photon and electrons from the valence band in insulators; this is called photo-conductivity. If the photon energy is smaller than the band gap, then electrons cannot be promoted and the photon is not absorbed: The material will be transparent to the electro-magnetic radiation of incident photon energy.[3] If the photon energy is greater than the energetic gap, then electrons will be promoted because of photon absorption. We will then observe that the material is opaque to electromagnetic radiation (figure 2.20). This also explains why metallic materials are usually opaque so in this instance the forbidden zone does not exist. This kind of explanation is qualitative and approximate because in materials partial absorption, elastic scattering, and so forth, can occur.

4. Electronic promotion can also happen by other direct or indirect mechanisms, for example, the excitation or deexcitation of other particles or elemental excitations like lattice vibrations (phonons). In all these transitions the usual quantities should be conserved. The

---

[3] $E_\gamma = h\nu$, where ν is the frequency.

transition is direct if, during electronic promotion from the valence band to the conduction band, no wave vector changes, as with GaAs. If the wave vector changes, the transition is indirect and there should be a mechanism in the transition that compensates this change. In some materials the minimum energy transitions are indirect. When this happens the band gap is indirect (figures 2.13 and 2.14).

The sole difference between an insulator and an *intrinsic semiconductor* is the band gap (table 2.3). The first one has a band gap larger than 2 or 3 eV; the intrinsic semiconductor has a much smaller band gap. This difference allows electronic promotion and the tunneling effect, stimulating the electrical conductivity of an intrinsic semiconductor owing to carriers in the conduction band. Indeed, thermal promotion of electrons in Si increases the number of electrons by a factor of 10^6 when we raise the temperature from 250 to 450 K. Nevertheless, for insulators the electronic promotion is much smaller.

Table 2.3 Band gaps

Insulators	Energy (eV)	Semiconductors	Energy (eV)
C (diamond)	5.33	Si	1.14
ZnO	3.2	Ge	0.67
AgCl	3.2	Te	0.33
CdS	2.42	InSb	0.23

Extrinsic semiconductors are ones in which chemical impurities or structural defects add (extrinsic) energetic levels to the band structure whose optoelectromagnetic properties are consequently modified by the new levels. The levels of interest to us are those that appear in the band gap. Static charge states are associated with extrinsic levels because they present a finite number of states.

Suppose that we have an impurity from a material chemically similar to the intrinsic semiconductor. Then the level lies near one of the bands and is called a *shallow level*. Consider the following two cases as examples.

1. We have Si (group IV) and we add a donor impurity, for example, P (group V). We then have an extra valence electron per atom compared to Si, and a (donor) extrinsic level will appear near the conduction band, which will be full at $T = 0$ K (figure 2.21).

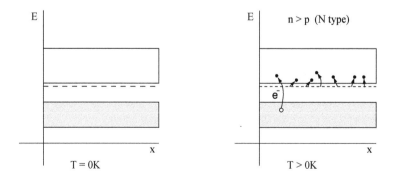

Figure 2.21 *N*-type extrinsic semiconductor.

At $T > 0$ K many electrons will be promoted from the extrinsic level (leaving in it a net static positive charge) and a few electrons will also come from the valence band so the electron quantity that contributes to conduction will be much greater than the hole quantity. Hence, $n > p$ and we will call this an *N-type semiconductor*. A *donor level* is defined as one that presents positive or neutral static charge states.

2. We have Si (group IV) and we add an acceptor impurity, for example, B (group III), so an (acceptor) extrinsic level will appear near the valence band and will be empty (figure 2.22). The conduction mechanism will be similar: Many electrons will be promoted from the valence band to the extrinsic level (giving in it a net static negative charge) and a few electrons will go to the conduction band. In this manner conduction is mostly done by holes ($p > n$): the *semiconductor* is *P type*. An *acceptor level* can be broadly defined as one that presents negative or neutral static charge states.

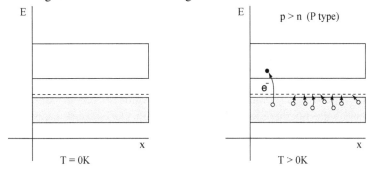

Figure 2.22 *P*-type extrinsic semiconductor.

Shallow levels, as we see, significantly affect conduction since they produce many more carriers of one type or another that are promoted into the respective bands. Yet optical properties are modified only for transition energies close to the preexisting ones or for low energies.

If an impurity is formed by a material chemically different from the lattice (e.g., Au in Si) *electron traps* appear from deep within the forbidden zone, far from the bands (figure 2.23). These levels, which may have negative, neutral, or positive static charge states, do not modify conduction properties because they only weakly affect the promotion of carriers. But optical properties change a lot because of the emergence of optical absorption lines that were not there before the change.

Note that in every one of our samples global electric neutrality exists because of the compensation between different types of mobile and static charges.

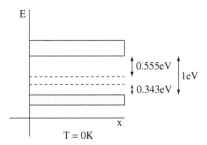

Figure 2.23 Electron traps (Au in Si).

2.6 CHARACTERIZATION OF CRYSTAL STRUCTURES

There is no fundamental difference between a solid structure and a molecular structure. In characterizing structures one finds a stable configuration of the nuclei and electrons that fits wave functions which represent the experimental characteristics—in particular, the distance between the crystalline planes and the orientations and positions of the atoms, ions, or molecules. Two key differences between a solid and a molecule structure are

- The large number of atoms implicated in the solid;

- Their regular distribution.

The most common experimental method for studying the structure of crystalline materials is *X-ray diffraction*. Appreciable diffraction is produced in a crystalline structure when it is illuminated with radiation of a wavelength comparable with the spacing between planes or smaller. In a typical crystalline lattice this spacing is usually of the order of 10^{-10} m, which obliges investigators to use X or γ rays.

When the crystal is illuminated with this radiation, the illumination beam deviates from rectilinear propagation because the crystal atoms or ions act as dispersion centers for the electromagnetic waves. (Sommerfeld defined diffraction as every deviation from rectilinear propagation that is not due to reflection or refraction.) Crystals are essentially similar to a three-dimensional diffraction lattice, where diffraction at each narrow slit is substituted by atomic dispersion.

The intensity pattern formed by the dispersed rays is the result of the interference of waves that have been dispersed by each atom in that direction and modulated by a characteristic factor of the dispersing atom. Then heavy atoms with many electrons cause dispersion with greater efficiency than do atoms with few electrons, owing to the density of the electronic cloud that surrounds them. When the crystal is comprised of more than one type of atom, each one scatters the incident radiation in a different way.

With these methods we can discern the bond lengths and angles between the molecules or ions that form a crystal and how they are bonded and ordered in it. When interpreting the spectrum of X-ray diffraction caused by crystals, we consider how the atoms or ions are ordered—a fundamental and difficult task for systematizing the results. This is why the study of how crystals are classified and ordered must precede the study of diffraction itself.

To understand the basis of these methods we consider first (as in section 2.3) a crystal order comprised of the same type of rigid spheres that correspond to an atom or ion. Now we want to simplify things so we have only one type of atom and only one atom in each unit cell. In analyzing the X-ray dispersion, imagine planes that are equally spaced and go through the layers of the crystal's atoms. In figure 2.24 several possible groups of parallel planes (e.g., a, b, and c planes) are shown for a cubic crystal. The groups of parallel planes differ from each other in their spacing and the density of dispersion centers.

Consider two parallel rays impinging at an angle θ on a group of equidistant planes separated by a distance d, as in figure 2.25. The maximum intensity occurs in the dispersed intensity when constructive interference has been accomplished (*Bragg's condition*):

$$2d \sin \theta = m\lambda,$$

where $m = 1, 2, 3, \ldots$ is an integer number called the *diffraction order*. This number is limited by the condition $\sin \theta = m\lambda/2d \leq 1$, λ being the wavelength of the incident radiation.

For rays like those in figure 2.25, Bragg's condition expresses the accumulation produced among all the rays that impinge at a given angle for all parallel planes, thus providing an in-

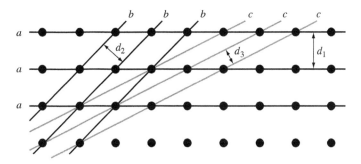

Figure 2.24 Groups of parallel planes in a simple cubic lattice in two dimensions.

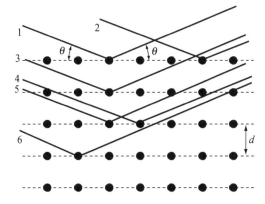

Figure 2.25 Dispersion of light incident at an angle θ on a group of equidistant parallel planes of a simple cubic lattice in two dimensions.

tense maximum whose structure is known as the *Laue pattern*. For a fixed distance between planes d and a fixed wavelength λ, the maximum of luminous intensity that corresponds to constructive interference alternates with positions of minimum luminous intensity that correspond to destructive interference. Thus, Bragg's condition can be used to determine the distance d in a particular crystal with a device known as a *crystal spectrometer*, as illustrated in figure 2.26.

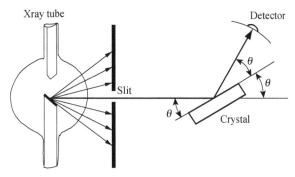

Figure 2.26 Crystal spectrometer, showing its principal components.

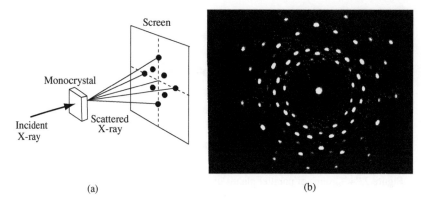

(a)   (b)

Figure 2.27 X-ray diffraction scheme of a monocrystal (a) and Laue pattern for the monocrystal SiO_2 (b).

When a monochromatic X-ray beam (i.e., a beam having only one frequency) is made to impinge on a simple crystal, a regular pattern is produced, as shown in figure 2.27. In this figure every point in the Laue pattern corresponds to the scattering direction for a family of planes.

If instead of a simple crystal the sample consists of a pulverized crystal containing many crystals randomly oriented, the directions corresponding to a family of planes that satisfy Bragg's condition will be distributed on conical surfaces around the direction of incidence (figure 2.28). If we have a photographic plate as the figure indicates, each conical surface produces a bright ring. This set of rings is known as the *Debye-Scherrer pattern*. When the produced patterns are analyzed, the internal crystal structure or X-ray wavelength can be determined.

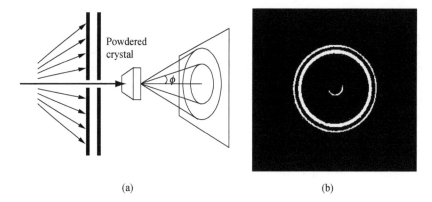

(a)   (b)

Figure 2.28 X-ray diffraction scheme of a powder (a) and the Debye-Scherrer pattern (b).

PROBLEMS

2.1. Demonstrate that $\vec{a}_1 = a\hat{\imath}$, $\vec{a}_2 = a\hat{\jmath}$, $\vec{a}_3 = (a/2)(\hat{\imath} + \hat{\jmath} + \hat{k})$ is a set of primitive vectors of a bcc lattice.

2.2. Demonstrate that $\vec{a}_1 = (a/2)(-\hat{i}+\hat{j}+\hat{k}), \vec{a}_2 = (a/2)(\hat{i}-\hat{j}+\hat{k}), \vec{a}_3 = (a/2)(\hat{i}+\hat{j}-\hat{k})$ is a set of primitive vectors of a bcc lattice.

2.3. Demonstrate that $\vec{a}_1 = (a/2)(\hat{j}+\hat{k}), \vec{a}_2 = (a/2)(\hat{i}+\hat{k}), \vec{a}_3 = (a/2)(\hat{i}+\hat{j})$ is a set of primitive vectors of a fcc lattice.

2.4. Demonstrate that $\vec{a}_1 = (a/2)(\hat{i}-\hat{k}), \vec{a}_2 = (a/2)(\hat{j}-\hat{k}), \vec{a}_3 = a\hat{k}$ is a set of primitive vectors of a fcc lattice.

2.5. Demonstrate that a simple hexagonal lattice is a Bravais one.

2.6. Demonstrate that the reciprocal lattice of the reciprocal lattice is itself.

2.7. Demonstrate that bcc and fcc lattices are mutually reciprocal.

2.8. Demonstrate that, if $V = \vec{a}_1 \cdot \vec{a}_2 \times \vec{a}_3$ is the volume of a primitive unit cell of a lattice, then the volume of the first Brillouin zone is $(2\pi)^3/V$.

2.9. Draw the first Brillouin zone for a simple hexagonal lattice.

2.10. Give an example of crystallographic direction, Miller indices, and equivalent sets in a simple cubic lattice. Interpret it.

2.11. Give an example of crystallographic direction, Miller indices, and equivalent sets in a bcc lattice. Interpret it.

2.12. Give an example of crystal direction, Miller indices, and both equivalent sets for the lattice with the following primitive vectors: $\vec{a}_1 = a\hat{i}, \vec{a}_2 = (3/5)a\hat{i} + (4/5)a\hat{j}$, $\vec{a}_3 = (4/5)a\hat{i} + (2/5)a\hat{j} + (1/2)a\hat{k}$. Interpret it.

2.13. CsCl crystallizes as a simple cubic lattice with the physical basis $\{0, (a/2)(\hat{i}+\hat{j}+\hat{k})\}$, in which the first element of the basis is Cs^+ and the second Cl^-. Give an example of the Miller indices of a plane containing both types of atoms. Give the Miller indices of two planes equivalent to the first one. What is the cesium ion fraction in that crystalline plane? Interpret it.

2.14. Give an example of crystal direction, Miller indices, and both equivalent sets for the lattice with the following primitive vectors: $\vec{a}_1 = a\hat{i}, \vec{a}_2 = b\hat{j}, \vec{a}_3 = (a/2)(\hat{i}+\hat{k}) + (b/2)\hat{j}$ a and b being different. Interpret it.

2.15. A monocrystalline sample of a specific material can be modeled by a Bravais lattice with the following primitive vectors: $\vec{a}_1 = a\hat{i}, \vec{a}_2 = (1/2)a\hat{i} + (\sqrt{3}/2)a\hat{j}, \vec{a}_3 = a\hat{k}$. Calculate the primitive vector set of the 2D Bravais lattice associated with the $(1\,1\,0)$ crystalline plane.

2.16. If the solid is amorphous, what do the energy bands look like?

2.17. The first-order spectrum of an X-ray beam diffracted by a NaCl crystal corresponds to an angle of $6°50'$ and the distance between planes is 2.81×10^{-10} m. Determine the wavelength of the X ray and the position of the second-order spectrum.

2.18. The first-order diffraction spectrum of an X-ray beam of $\lambda = 0.7093$ Å corresponding to the planes $\hat{\imath} + \hat{\jmath} + \hat{k}$ of a fcc crystal of Ni takes place at an angle of 10.04°. Find the corresponding distance between planes. How many diffraction orders of this kind of plane can we observe?

2.19. An X-ray beam of $\lambda = 2.2897$ Å that impinges on planes $\hat{\imath} + \hat{\jmath} + \hat{k}$ of a fcc crystal of Cu provides only one nontrivial diffraction peak corresponding to an angle of 33.27°. Find out the corresponding distance between planes. What is the diffraction order corresponding to an angle of 58.179° if the incident wavelength is $\lambda = 0.7093$ Å?

Chapter Three

Imperfections

In chapter 2 we supposed that we had perfect crystalline solids. This is an idealization because, in the first place, solids are not infinite; we have to take into account the effects that external surfaces may have. Our supposition is an idealization also because solids can be at temperatures greater than zero; hence the positions of atoms, ions, and molecules will not be the ideal crystalline equilibrium positions. On the contrary, they will vibrate about these equilibrium positions. So we will not dwell on ideal crystalline solids anymore. Besides, it is difficult, even with these restrictions, to obtain a perfect monocrystalline solid, as we already noted.

3.1 DEFECTS: TYPES OF DEFECTS AND VIBRATIONS

Having a finite solid is not a defect since the macroscopic lengths of the solid sample are usually greater (by many orders of magnitude) than its interatomic distances; therefore we can say that we have a *perfect* lattice of infinite size—although we have appreciable surface effects (see chapter 11). By contrast, vibrations are not a defect because they are always present at temperatures higher than absolute zero; hence defects are only differences with regard to a perfect finite crystal. These differences, then, can be classified according to their dimension: *point defects* (0D), *dislocations* (1D), and *planar defects* (2D). Another equivalent classification indicates whether the imperfect region at the atomic scale is enclosed in one, two, or three dimensions (planar defects, dislocations, and point defects, respectively).

Imagine that atomic vibrations at lattice sites behave like billiard balls at the sites joined by elastic springs (see figure 5.3). Their heightened vibrations can be understood from their thermal rattling (caused by the temperature). When the vibrations are quantized one can detect their contribution, for example, to the heat. In a similar fashion, electrons in excited states deviate from the perfect crystalline lattice.

3.2 POINT DEFECTS: FRENKEL AND SCHOTTKY DEFECTS

Point defects can be classified as follows:

- *Vacancies*: There is a lattice site without a component.

- *Interstitial atoms*: A component that is not the most common in the solid and is called the solute lies at a point that is not a lattice site.

- *Self-interstitial atoms*. Interstitial atoms of the most common component in a solid, called the solvent.

- *Substitutional atoms*. There is a component at a lattice site different from the one it was supposed to be.

Before embarking on our main considerations, we must note that the smaller the solute atoms, the larger is the probability of having interstitial atoms and the smaller that of having substitutional atoms. For this reason the most significant atoms for interstitial defects are C, N, and O (all with a radius less than 0.8 Å).

Component vibrations in an ideal lattice are a source of defects. If an atom vibrates with enough amplitude, it can leave a vacancy that simultaneously occupies another one (vacancy diffusion), or occupies an interstitial position (a combination called a *Frenkel defect*), or occupies a superficial position (a combination called a *Schottky defect*).

The equilibrium concentration of each type of point defect depends on the temperature. We are going to examine the free-energy variation in the quasistatic isothermal and isobaric creation of a defect: $\Delta G = \Delta H - T \Delta S$. The enthalpy variation ΔH will be positive and is the thermal energy to move an atom at constant pressure ($E_D \sim 1$ eV/atom). As with isothermal and isobaric processes, where we require that the free-energy variation is null, it follows that the entropy variation is greater than zero. Consequently, for a positive temperature we will have more disorder. At equilibrium, we have Schottky defect concentrations according to the Boltzmann distribution, which is approximately

$$\frac{n_e}{N} = A e^{-E_D/k_B T},$$

generally taking $A \approx 1$.

The number of Schottky defects radically increases an insulator's conductivity. Meanwhile, for semiconductors the electrical conductivity dependence on temperature is due to the thermal excitation of electrons from the valence band to the conduction band; in insulators this excitation is much less and the induced electric conductivity is tiny. Nevertheless, Schottky defects (vacancies) increase electrical conductivity because ion diffusion through vacancies is much easier than ion diffusion (induced by an electric field) in a lattice without those vacancies. This is confirmed in electrodes, because the corresponding ions appear. It can also be indirectly verified. For example, if we have NaCl and introduce Ca^{++} impurities, the impurities cause creation of vacancies at Na^+ sites to achieve neutrality in the total charge. Conduction rises in proportion to the calcium introduced. A similar check can be done on the variation of defect concentrations with temperature, but here the diffusion constant changes too, so the vacancy concentration effect is not readily distinguishable.

It follows that, when there is a vacancy of one of the crystal components, the preservation mechanism of the electric charge shows up in interstitials of the same component, or in the appearance of vacancies of another component that can compensate the charge in the first vacancies. This scheme conserves charge by means of a localized electron in the point defect's neighborhood, where the charge changes. This bounded electron has a discrete spectrum of energy levels. Excitations between these levels produce optical absorption lines similar to the ones created by isolated atoms in the forbidden optical zone (between $\hbar \omega_T$ and $\hbar \omega_L$). This kind of electron-defect structure is called a *color center* because it confers a strong coloration on crystals that were initially colorless.

Color centers have been much studied in alkali halides, which can be colored by exposure to X or γ rays—which produces vacancies caused by collisions with highly energetic photons—or by heating the crystal in an alkali vapor. In this instance alkali atoms are introduced into the material, as the chemical analysis shows. But the density decreases in proportion to the excess concentration of alkali atoms; hence, they are not interstitially absorbed (figure 3.1) and spare electrons are bounded to halide vacancies. This effect is confirmed by the observation that the optical absorption spectrum does not depend on the

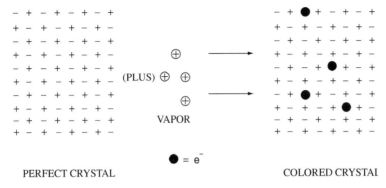

Figure 3.1 Coloration process of an alkali halide.

kind of alkali atom used to heat the crystal; the only effect, then, is to produce negative vacancies bound to electrons.

We give the name F center[1] to an electron bound to a negative vacancy. Its energetic levels reproduce the main characteristics of an isolated atom. The only difference is that the crystal field imposes symmetries different from the spherical one to the electron. In addition, when the crystal is deformed we can modify the imposed symmetries by making the crystal field modify the spectrum. In this sense, if a neighbor of an F center is a substitutional atom, the symmetries are drastically reduced—and then it is called an F_A center.

The absorption spectrum structure of colored crystals is not so simple because the F centers are not the only color centers or lattice electrons localized around vacancies. We can have more complex centers, such as the M center (in which two contiguous negative vacancies bind two electrons) or the R center (in which three contiguous negative vacancies bind three electrons).

We could regard holes that produce bound systems as analogous to the preceding defects. Yet they have not been observed. Only a V_K center has been observed, where a hole binds two negative lattice ions in the absence of defects, or an H center where a hole binds a negative lattice ion to a negative interstitial ion. In both instances, pseudomolecules are created (similar to Cl_2) whose spectra are like molecular ones. It is not possible to observe compounds of atoms with positive oxidation numbers, where an electron serves as a bond, since there are no covalent molecules of Na, for example.

The absorption linewidth of the color centers is much larger than that of isolated atoms—which are narrow because they can decay only by emitting electromagnetic radiation and thus have long decay times—since a color center, which is strongly coupled with the crystal, can decay by way of lattice vibrations.

In ionic crystals (insulators), when an electron is promoted to the valence band, it can move in a localized form with charge screening due to the lattice's local deformation, which is caused by the intensity of electron-ion interaction in this type of crystal. Thus, the effective mass of this electron is different from the effective mass predicted by pure band theory. The elemental excitation formed by an electron and the associated local deformation field of the lattice is known as the *polaron*.

---

[1]From the German *Farbzentrum*: color center.

The most typical point defects are the ones that we have seen so far. Another kind of defect, called the *exciton*, is based on the fact that perfect crystal ions are in excited electronic states. We can express this more precisely by stating that an exciton is a bound state formed by an electron-hole pair. This pair can transport energy but not charge; it is electrically neutral. The binding energy range of an exciton is between 1 meV and 1 eV (table 3.1). This makes the exciton formation energy minor with respect to the direct or indirect band gap energy. We can measure the binding energy by means of optical transitions from the valence band because of the difference between the energy needed to create an exciton and the energy needed to generate a hole and an electron that are both free, and also by comparing these energies in recombination luminescence or in exciton photoionization where a high exciton concentration is needed.

Table 3.1 Binding energy for excitons (in meV)

Si	14.7	BaO	56	RbCl	440	GaAs	4.2	InSb	~ 0.4
Ge	4.15	InP	4.0	LiF	~ 1000	AgBr	20	GaP	3.5
KI	480	AgCl	30	CdS	29	KCl	400	TlCl	11

The two extreme cases are, first, Frenkel excitons (strongly bound excitons) that are not localized as a whole—they can be everywhere in the lattice—but describe a small orbit, as though they were atoms; second, the weakly bound excitons or Mott-Wannier excitons that describe large orbits compared with their interatomic distance (figure 3.2).

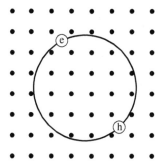

Figure 3.2 Mott-Wannier exciton, an electron-hole pair weakly bound and free to move in the solid.

The exciton gas with weak interactions can condense at low temperatures into electron-hole drops when the exciton concentration is large.

The time it takes to form an exciton is of the order of 1ns; the exciton recombines in roughly 8 μs. But the electron-hole drops decay in 40 μs and increase a lot, for example, in stressed Ge. Inside a drop the excitons dissolve into a degenerate Fermi gas of electrons and holes with metallic properties.

The width of the recombination lines in electron-hole drops is due to the kinetic energy distribution of the electrons and holes, in contrast to the sharper recombination lines of free excitons (figure 3.3).

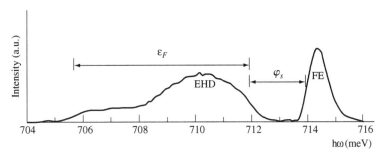

Figure 3.3 Recombination spectra of free excitons (FE) and electron-hole drops (EHD) in Ge at 2.04 K. E_F is the Fermi energy in the drop, and φ_s is the drop's cohesion energy with respect to the free-exciton cohesion energy.

3.3 DISLOCATIONS AND DISCLINATIONS

In 0D defects (point defects) the most obvious effect in crystalline solids is the appearance of new optical absorption properties. In unidimensional defects (dislocations), the most evident effect resides in the order of magnitude change in force needed to deform a crystalline solid. Yet if the number of defects increases very much, the force starts to rise. These are the main effects of dislocations, and this is the phenomenology that interests us.

First, one must distinguish between the two principal types of dislocations: *edge dislocations* and *screw dislocations*. They are illustrated in figure 3.4. Dislocations will not always be straight, so in general a dislocation is a unidimensional region inside the crystal that has the following properties:

- Far from the dislocation the crystal is almost perfect (locally),

- Around the dislocation the atomic positions are different from the perfect crystal positions, and

- A non-null *Burgers vector* does exist.

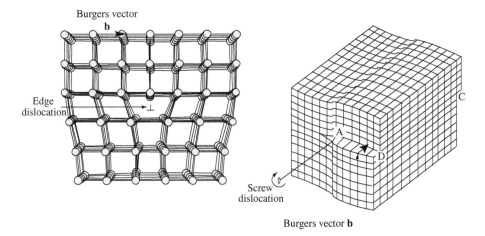

Figure 3.4 Edge and screw dislocations.

To apply this definition we have to define the Burgers vector, given that we have a closed curve in a perfect crystal that goes through a succession of lattice sites—and therefore is a sequence of displacements in the Bravais lattice. Then the same sequence in the preassumed dislocated crystal is covered. This circuit must be covered far from the pre-assumed dislocation to avoid ambiguities about the meaning of "same sequence." If we do not succeed in reaching the initial point, then the curve has surrounded a dislocation; so the remaining way to achieve the initial point defines the Burgers vector. With this perspective we can determine the differences between edge and screw dislocations. In an ideal edge dislocation the Burgers vector is perpendicular to the dislocation line, and in a screw dislocation the Burgers vector is parallel to the dislocation line.

With a perfect crystal it is difficult to displace one whole atomic plane over another. The required force to be overcome is proportional to the number of atoms in that plane. Nevertheless, if there is a dislocation, moving an atomic plane is equivalent to moving the dislocation (*caterpillar effect*) alone (figure 3.5). If the number of defects increases a lot, once again it will be difficult to move the plane because the dislocation effects can cancel out, including the possibility that the dislocations themselves could be canceled out. Or, if there is some other kind of defect (interstitial atoms, vacancies, planar defects, etc.), they pin dislocations at points of the lattice and make it more difficult to move the plane.

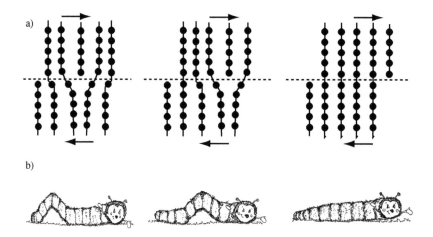

Figure 3.5 Dislocation movement. In (a) the dislocation slips in the solid until it finally disappears at its surface. In (b) the analogy of a caterpillar's movement is demonstrated.

A related subject is when we have a sample of a malleable and ductile metallic material. If we fold it several times and restore it to its initial position, we can do so only a finite number of times. This is because every time we fold the sample we introduce defects until there are so many that the sample hardens, does not admit more plastic deformation, and finally breaks.

Dislocations, then, change material properties, which is why this question has become the subject of utmost interest. Also, it is significant that dislocations—mainly screw dislocations—favor the growth of monocrystalline materials around the same spiral form (figure 3.6). As a consequence, the sample has only one defect, a screw dislocation, because the sample's interaction with atoms, molecules, or ions is greatest at the screw dislocation.

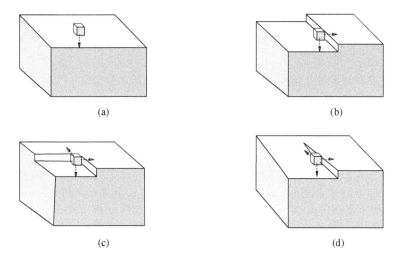

Figure 3.6 In (a) atoms are weakly attracted by a plane, but in (b) an edge attracts them more strongly, and a corner even more so in (c). If the crystal has a screw dislocation, this attracts atoms in an optimum way, as shown in (d).

Since an edge dislocation can be visualized in a perfect crystal whose atoms in one crystalline semiplane have been removed and whose edges have been stuck together as the line bounding the semiplane, a *disclination* can be visualized as a line that bounds the cylindrical sector in a perfect crystal, where the edges have been stuck together again.

3.4 PLANAR DEFECTS AND GRAIN SIZE

Planar defects are grain boundaries that have two directions and separate regions of different orientations or crystalline structures. They include external surfaces (where the solid ends, which makes the surfaces important because atoms there are in a higher-energy state than the atoms in the volume), grain boundaries (mentioned when we introduced polycrystalline materials), pile defects (they occur in planes of compact structures) or phase boundaries (interphases, for instance, of different phases of the same or different atomic species). As noted, planar defects in the broad sense show shape anisotropy in materials because every planar defect defines a surface, and every surface that bounds or separates a volume by a shape yields anisotropies.

For grain boundaries we can distinguish the inclination boundaries where the angle of contiguous grains (called the disorientation angle) between crystallographic directions is small and can be modeled as a sequence of dislocations. The other distinguished grain boundary is the twin crystal boundary which defines a specular symmetry plane. Twin crystals can be deformation ones—when shear forces are applied to metallic materials with bcc and cph structures—or induced by heat, which are typical of fcc structures.

The grain size is the crystal volume confined by grain boundaries. To improve material properties one might increase or decrease the mean grain size. For example, the phenomenon of grain growth could be preceded by a restoring stage (stress is released by dislocation movement) and a recrystallization stage (where new grains are created with a low density of defects). The malleability and ductility of a material can thereby be raised.

Also a material can be hardened by grain size reduction because the grain boundaries—not the inclination ones—anchor dislocations and create a discontinuity in the slip planes.

3.5 DEFECT DETECTION

In the first place the direct methods of defect detection are simpler for planar defects than they are for dislocations, and they are simpler than for point defects because of their dimension. It is easy to observe grain boundaries in crystalline materials that are anisotropic. This anisotropy is apparent when we cut and polish the sample and attack it with a reagent; the sample acts differently according to its exposed crystallographic directions. Hence the resulting *roughness* will cause the received light to reflect upon dispersion. For example, if observed through an optical microscope, the different material grains will appear in a distinct form. Reflection electron microscopy, in particular scanning electron microscopy (SEM), is commonly employed to study inclination boundaries, as with X-radiation diffraction.

Sample dislocations can be observed by electron microscopy. These techniques admit much larger augmentation, and the transmission technique (TEM) is the one most used, although narrow samples are required to prevent electrons from the optical system from being absorbed by the sample. Its main advantage is that, besides permitting many augmentations, linear defects inside the solid can be seen. Also, a dislocation can be decorated with a material that is chemically affine because during the dislocation interatomic distances are modified and therefore the dislocation has forces not locally compensated by the solid.

Point defects can be observed by electron microscopy techniques (by tunneling and atomic force, among others). All these techniques are direct, although many indirect techniques can measure how many types of defect are present in a crystalline solid.

3.6 AMORPHOUS MATERIALS

Crystalline solids, as we previously explained, might present imperfections that cause an imperfect crystalline lattice. If the number of imperfections is high, the solid can no longer be considered crystalline because its defined order is not long range. In this case the solid is called a glassy material if it has short-range order or amorphous if it has no order.

Generally, solids (i.e., materials that define a volume and a shape) have a crystalline (or quasicrystalline) structure in which the solid energy is a minimum (this does not occur, for instance, when the frustration phenomenon appears; see section 6.2). Thus, amorphous or glassy materials are not in minimum energy states and are therefore in unstable or metastable states that evolve in times that are rather long toward their crystalline state of minimum energy. This leads to amorphous materials that are extremely viscous liquids, which can change to their stable liquid versions if we raise the sample temperature until the heat that we apply to the sample's crystalline version makes it melt. Between the glassy transition temperature, where flow times are long compared with stable liquid ones, and the melting temperature, we note every intermediate viscosity for which we can consider the glassy material as a supercooled liquid.

Amorphous materials provide X-ray diffraction spectra similar to liquid ones. Dark and light circular zones correspond to diffraction minima and maxima. If we reconstruct from these diffraction figures the average density before a particular distance, we find the mean bond lengths of the glassy structure (figure 3.7).

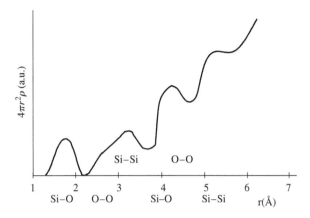

Figure 3.7 Average density of glassy SiO_2 as a function of interatomic distance.

Noncrystalline materials will be studied thoroughly in chapter 9.

PROBLEMS

3.1. What is the typical concentration of Schottky defects at room temperature?

3.2. The concentrations of Schottky defects in equilibrium for copper are 4×10^{-6} and 15×10^{-5} at 1000 and 1325 K, respectively. What is the formation energy of these defects?

3.3. Calculate the formation energy E_D of Schottky defects ($n/N = e^{-E_D/k_B T}$) in aluminum when you know that the number of defects at 773 K is 7.57×10^{23} m^{-3} and that the aluminum density at this temperature is 2.62 g $\cdot$ cm^{-3}.

3.4. What is the difference between the Burgers vector in an edge and a screw dislocation?

3.5. Calculate the disorientation angle of an inclination boundary as a function of the separation between consecutive edge dislocations and their Burgers vectors.

3.6. Discuss what would happen if we had a sequence of screw dislocations instead of edge dislocations.

3.7. Figure out the possible mechanisms of substitutional and interstitial diffusion in solids.

3.8. A biphasic material has precipitated particles of radius r_i and concentration c_i. Knowing that the energy per unit of area of the grain boundary is ϵ_0, calculate the total reduction of a grain boundary's energy per unit of volume when a thermal process makes it possible to increase the radius to r_f and reduce the concentration to c_f.

3.9. Knowing that the formation energy of a dislocation is proportional to the square of the Burgers vector modulus, discuss the necessary conditions for splitting a dislocation into two.

3.10. Discuss the following assertion: Nitrogen and oxygen can occupy interstitial positions *because* they are gases.

Chapter Four

Electrical Properties

This chapter is about the electrical properties of metallic and semiconducting materials. Here we leave out the investigation of these properties in insulators and superconductors, which we treat in subsequent chapters.

Since materials can be considered in the framework of band theory—at least the crystalline ones—it seems reasonable to use this theory to describe the electrical properties of these materials. Nevertheless, obtaining results often implies long and complex mathematical developments, which could be avoided by using semiclassical or classical theories that yield satisfactory results. That is why in the first two sections of this chapter we perform the following simplifications (*Drude's model*):

- The atoms in a conductor are regarded as structured into the nucleus and its inner electrons (ions), and the conduction electrons, which are considered free in the solid and behave as a gas: the electron gas, also called the Fermi gas. We conceptually separate the ionic effect from the conduction electron effect, which implies that the description will not be purely a quantum one.

- Between collisions there are few interactions of one electron with others and with ions; these are called, respectively, the independent-electron approximation (a good one) and the free-electron approximation.

- Collision between electrons is an instantaneous process.

- The mean time between collisions is τ (relaxation time).

- The electrons achieve thermal equilibrium uniquely by means of collisions.

In section 4.3 we use concepts from band theory that were already introduced in chapter 2.

4.1 ELECTRICAL CONDUCTIVITY AND TEMPERATURE

The *electrical conductivity* σ, or simply conductivity, is a physical property that classifies materials into insulators or dielectrics, semiconductors, semimetals, conductors or metallic materials, and superconductors.

In this chapter we focus on conductors and semiconductors (and semimetals). In chapter 6 dielectrics are examined, and superconductors are treated in chapter 7. Hence, we view here materials with a large, but not infinite, conductivity that decreases when the temperature rises owing to the augmentation of lattice vibrations. Such is the situation for conductors. In addition, we look at materials with a small, but not null, conductivity that increases or decreases when the temperature rises, as with semiconductors or semimetals, due to the heightened electronic promotion in the first (section 2.5) and to the additional collisions with lattice vibrations in the second.

For materials with low conductivity (insulators), the conductivity increases when the temperature is raised because of vacancy diffusion (section 3.2) and, as a minor cause, electrons are promoted to conduction bands. Superconductors—which have practically infinite conductivity independent of lattice vibrations—lose their superconductor behavior at a critical temperature (see chapter 7).

We know that the conductivity σ is defined as $\vec{j} = \sigma \vec{E}$, so the resistivity is its inverse. If we apply a constant and homogeneous electrical field to a conductor, we will have an electron that accelerates between one collision and the next with a magnitude $\vec{a} = -e\vec{E}/m$, so its velocity at time t will be $\vec{v}(t) = \vec{v}_0 - (e\vec{E}/m)t$, $\vec{v}_0$ being its velocity just after the last collision. As the velocity $\vec{v}_0$ has a random direction, the average velocity $\langle \vec{v} \rangle = -(e\vec{E}/m)\tau$. At the same time, the electron flow $\vec{j}$ can be related to the calculated average velocity: $\vec{j} = -ne\langle \vec{v} \rangle$. Therefore,

$$\sigma = \frac{ne^2\tau}{m}.$$

Here the only unknown parameter besides the one that can depend on temperature is τ (also called the *mean free time*). This parameter is inverse to the probability density of a collision, which is proportional to the number of electrons (almost constant in metallic materials) and to the number of vibrations of the lattice sites (which increases with temperature). Therefore, the conductivity is reduced when the temperature rises. At low temperatures, the resistivity falls to low values (different from zero in the absence of superconductivity). This resistivity saturation appears because, although the phonons (quantized lattice vibrations) decrease to zero at low temperatures, there are electronic collisions with crystalline defects, impurities, and so on that convert the density of imperfections into a control parameter with which we can increase resistivity according to our criteria.

Electrical conductivity plays a major role in explaining the thermal conductivity for metallic materials, as we see in chapter 5.

4.2 THERMOELECTRIC AND GALVANOMAGNETIC EFFECTS

Every thermoelectric effect comes from the experimental fact that in materials electrical stimuli provoke thermal responses and thermal stimuli provoke electrical responses. The whole process operates in a reversible way, which differs, for example, from the Joule effect, indicating the power dissipated into heat when an electrical current passes by. This is because even in isotropic materials, for instance, a temperature gradient is accompanied by an electrical field opposed to the gradient. This phenomenon, called the *Seebeck effect*, can be modeled by the equation

$$\vec{E} = Q\nabla T,$$

where the proportionality constant Q must be found, assuming that in the stationary state the electron velocity produced by the temperature gradient is equal and opposite to the one caused by the electrical field. The electrical field velocity is $v_E = -eE\tau/m$ and the thermal gradient velocity is $v_Q = -(\tau/6)v \, dv/dx$ where the factor of 6 is due to the existence of six movement directions in a tridimensional space. If $dv/dx = (1/2v)(dv^2/dT)dT/dx$, then

$$Q = -\frac{1}{3e}\frac{d}{dT}\frac{mv^2}{2},$$

which is proportional to the variation of the electron kinetic energy with respect to temperature. In some materials the sign of Q changes because sometimes the carriers are not electrons but holes. The Seebeck effect is highly useful when it is employed for two different materials (figure 4.1), since that is when an electrical field can manifest in a stationary state in metallic materials. If we take this into account, we can conveniently calibrate the effect of a *thermocouple* on the relation between temperature difference and electrical potential difference; what is more, it can serve as a precision differential thermometer.

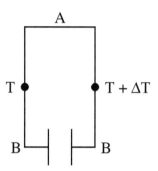

Figure 4.1 Scheme of the Seebeck effect, assuming that we have two different materials A and B. The junction points between the two materials are called reference points or welds and test points or welds. The induced electrical field accumulates charge in the capacitor.

A thermoelectric effect of similar nature is the *Peltier effect*. This one indicates that a charge flow, or an electrical current, through the junction between two different materials having the same constant temperature is accompanied by heat absorption and emission; that is, a charge flow provokes a heat flow. This heat flow is proportional to the differential thermoelectric power difference between the two materials. Since conductors have similar differential thermoelectric powers, it follows that the initial effect is small. In spite of this, in 1838 Lenz had already used Sb and Bi and reached values of electrical current nine times greater than in conductors. Nowadays, in applications where extrinsic semiconductor alloys are used, as, for example, $Si_{0.78}Ge_{0.22}$, the Peltier effect can reach values almost 50 times greater than the values corresponding to conductors. So Peltier cells (figure 4.2) can be made that benefit from the thermal energy absorption at the junction and cool a ceramic plate in contact with the system. With this device portable refrigerators and similar systems can be made.

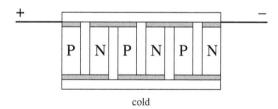

Figure 4.2 Schematic figure of a Peltier cell. The plates that bound the semiconductors are made of a ceramic material.

The continuous Peltier effect can also be considered: It consists of the Peltier effect in a material that continuously changes its composition.

Finally, we must comment on the *Thompson effect*, which is the response to the coupling of an electrical current and a temperature gradient in the system. This provides heat flows in the system that are unexplained by the thermoelectric effects described above.

In addition to these thermoelectric effects for isotropic materials, other effects exist where the anisotropic nature of the materials plays an important role.

Galvanomagnetic effects are the kinetic phenomena that happen when magnetic and electrical fields act simultaneously. The best-known phenomenon is the *Hall effect*, which provides the majority carriers in the material. In figure 4.3 this situation is drawn for electrons as carriers. The magnetic field, through the *Lorentz force* (see chapter 6), causes carriers to deviate in such a way that there is a lateral charge accumulation that produces a transverse electrical field. In addition, as some of the carriers go to the sides, in principle there are fewer available to favor a longitudinal current.

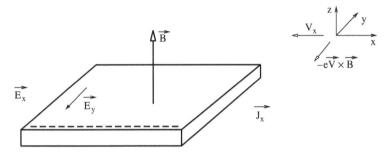

Figure 4.3 Scheme of a device to study the Hall effect with electron carriers.

So we define the *magnetoresistence* as $\rho(B) = E_x/J_x$; hence, as we see that the transverse field E_y in the stationary state must be proportional to B and to J_x (Lorentz law), we define the *Hall coefficient* R_H as

$$R(H) \equiv \frac{E_y}{J_x B}.$$

Using the Lorentz formula, we obtain the following: When the majority carriers are electrons the transverse field is $E_y = -(B/ne)J_x$, and so $R_H = -1/ne$, which does not depend on the conductor except for the carrier density n (problem 4.3). From the Lorentz force one discovers that, if the carriers are holes and their charge is positive, then the direction of the longitudinal velocity changes, the force is in the same direction, and the majority carriers accumulate in the same place as do electrons (figure 4.4); therefore, the Hall coefficient can be expressed in a general way as

$$R_H = \frac{1}{nq},$$

where q is the carrier charge, which makes the effective mass positive in order to be able to apply the classical ideas of force, acceleration, and so forth. In this way, by measuring the direction of the transverse electrical field, we can identify the majority charge carriers and their density.

It is worth mentioning thermogalvanomagnetic effects, which reveal the interaction of magnetic fields and thermoelectric effects. For example, in the *Nernst effect* a magnetic field modifies the Seebeck effect, in the *Ettingshausen effect* a magnetic field modifies the Peltier effect, and in the *Leduc-Riggi effect* a magnetic field modifies thermal conductivity.

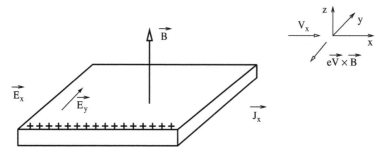

Figure 4.4 Scheme of a device to study the Hall effect with hole carriers.

4.3 SEMICONDUCTOR ELECTRONIC DEVICES

The *P*-*N* Junction

The *P*-*N junction* is the name for a piece of a P-type extrinsic semiconductor joined together with a piece of *N*-type extrinsic semiconductor. From a practical viewpoint, let us take an extrinsic silicon monocrystal into which we introduce donor and acceptor impurities in different zones of the solid.

If we imagine a *P*-*N* junction, in the *P* zone there is a concentration of free holes moving as a gas, and in the *N* zone there is a concentration of free electrons. Since the two types of carriers are active, the sort of devices made up of this type of junction are called *bipolar devices*.[1] Diodes are devices that let the electrical current pass in one direction but not its opposite. These are semiconductor diodes, but also other types, for example the metal-insulator-metal (MIM), which have a wide response frequency band and are very fast. As with normal gases, the two components tend to diffuse into the other zone; and since they are complementary, they recombine. Therefore, a *depletion layer* of free carriers occurs, showing a net charge density (due to the underlying ions). In the *N* zone it will be positive and in the *P* zone it will be negative. This net charge density will create an electrical field from the *N* zone to the *P* zone that inhibits the diffusion of more carriers to the other zone. Consequently, in the absence of an external electrical field, the electrical potential is as shown, idealized in figure 4.5.

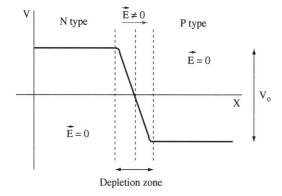

Figure 4.5 Idealized electrical voltage in a *P*-*N* junction without an external field.

---

[1] Concretely, a P-N junction is a *diode*.

For any semiconductor in an equilibrium situation,

$$np = M_c M_v e^{-E_g/k_B T} = n_i^2,$$

where n_i is the concentration of electrons in the intrinsic semiconductor and M_c and M_v are the equivalent densities of states in the conduction and valence bands, respectively. This expression is not valid when there are nonequilibrium mechanisms generating carriers, as for example when illumination with photons of higher energy than E_g creates intense electrical fields or when there is carrier injection through the semiconductor junction. If we have an internal voltage in a P-N junction, then $N_A N_D \approx np = n_i^2 e^{eV_0/k_B T}$ and therefore we obtain

$$eV_0 = k_B T \ln \left(\frac{N_A N_D}{n_i^2} \right),$$

where N_A and N_D are the densities for acceptors and donors, respectively. For silicon, with typical values of impurities, we obtain for $n_i = 6.3 \times 10^{15}$ m^{-3}, $T = 300$ K, and $N_A \sim N_D \approx 10^{22}$ m^{-3} a value of $V_0 = 0.7$ V. The energy bands are shown in figure 4.6.

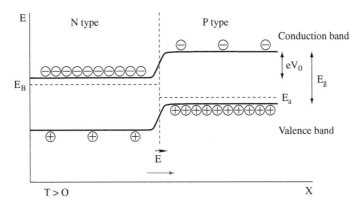

Figure 4.6 Energy bands in an unbiased (without an external voltage) P-N junction.

As you can see, the electrical field in the depletion layer does not allow carriers to cross from one zone to the other; hence, there is no current. Nevertheless, if we apply a forward-bias voltage V (i.e., the electrical field goes from the P zone to the N zone) we reduce the internal field by helping electrons in the conduction band and holes in the valence band to conduct, thereby producing a current that ends in the battery (figure 4.7).

Hence, the I-V characteristic of the diode has an exponential form:

$$I = I_s \left(e^{eV/K_B T} - 1 \right),$$

where $I_s \sim 1$ μA is the reverse current obtained when a reverse-bias voltage is imposed, $V < 0$, and it is proportional to n_i^2. For a large reverse-bias voltage ($|V| > V_R$, where V_R is the breakdown voltage), the intensity across the diode grows tremendously due to the existence of conduction by the *tunneling effect* (if $e|V_0 + V| > E_g$, where E_g is the gap energy) or the *avalanche effect*, or both.

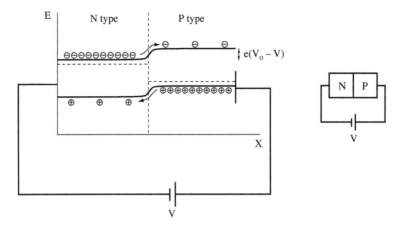

Figure 4.7 Energy bands in a biased (with an external voltage) *P*-*N* junction with a forward-bias voltage.

Transistors and Field Effect Transistors

A semiconductor (bipolar) *transistor* (transfer resistor) or *BJT* (bipolar junction transistor) has two *P*-*N* junctions (figure 4.8). If we observe its energy bands when an external voltage has been applied to the system such that the base-emitter junction is forward biased and the base-collector junction is reverse biased, the transistor is biased in the linear active zone, as in figure 4.9.

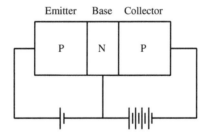

Figure 4.8 Scheme of a *P*-*N*-*P* bipolar transistor biased in the linear active zone.

With a small $V_E - V_B > 0$ a large quantity of carriers is injected into the base from the emitter. These carriers will recombine quickly and will go across the base ($I_B \approx I_E$). Yet transistors are made with such a thin base that the carriers injected into the base are in the collector before we realize it. Therefore, we obtain $I_B \approx 0$ and $I_E \approx I_C$. Because $V_B - V_C \gg V_E - V_B$, the power of the signal that comes from the collector is larger than the one that enters the emitter, and hence transistors behave as power amplifiers. This is one of the most important analog applications of the BJT biased in the linear active zone. Bipolar junction transistors that have been differently biased have their most significant applications in digital electronics.

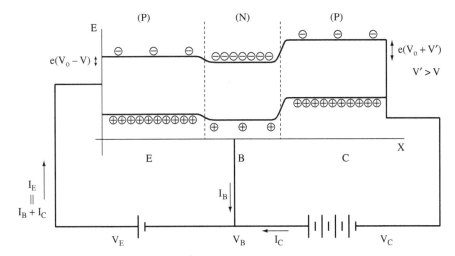

Figure 4.9 Energy bands in a P-N-P bipolar transistor biased in the linear active zone.

We have noted that the junction transistor is a bipolar device. Even so, we can fabricate a FET (field effect transistor), a device where only one type of carrier conducts.

The FET can be a JFET (junction field effect transistor) or an IGFET (insulated gate field effect transistor); the MISFET (metal-insulator-semiconductor field effect transistor) (figure 4.10) and the MOSFET (metal-oxide-semiconductor field effect transistor) belong to this last type. In the FET the gate contact generates an electrical field (the name FET comes from this) as does a planar capacitor, and it biases carriers into N and P so the available carriers of one type are more numerous than usual (figure 4.11). Then, by setting a slight voltage between the source and the drain contacts, we can produce intensity that bypasses the channel linking the source and the drain—called the N or P channel, depending on what type of semiconductor it is made of—and makes the system much more intense than it would be if a field through the gate had not been set.

Finally, if bipolar and field effect technologies are combined with insulated gates then the IGBT (insulated gate bipolar transistor) can be made, which is useful for high-power applications that achieve gains (the ratio between the power leaving the device and the one that controls the gate) up to 10^7.

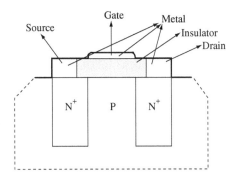

Figure 4.10 Scheme of a MISFET with an unbiased N channel.

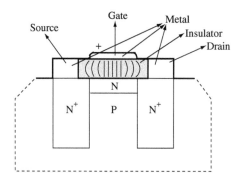

Figure 4.11 Scheme of a MISFET with a biased N channel.

Semiconductor Processing

Semiconductor processing consists of transforming silicon into semiconductor materials suitable for microelectronic usage in *chips*. It can be summed up in the following steps.

1. *Purification.* After producing 98% pure Si by reducing SiO_2 with C, it is purified at 300°C with HCl, forming $SiHCl_3$ as well as other compounds with Cl. The $SiHCl_3$ is liquid at $T < 31.8$°C and can be separated by fractional distillation. Thus we obtain a concentration of impurities less than 1 ppb. To obtain pure electronic Si, deposition by reduction is produced by using pure H_2 over Si at a temperature of the $SiHCl_3$ vapor of 1000°C. The by-products can be recycled to obtain more Si. The only difficulty with this process is its high energy consumption. There are other ways to purify Si, such as, for example, the pyrolysis of SiH_4.

2. *Crystal growth.* Silicon is melted and a crystalline seed is introduced with the desired orientation (growth plane parallel to the surface) into the liquid Si. The temperature of the liquid Si is lowered until it starts to solidify. At this moment the sample is slowly pulled so more crystal planes are solidified. At the same time the sample is rotated to help the solidification (figure 3.6).

3. *Doping.* An extrinsic substrate of Si is necessary to fabricate chips where a layer of a different extrinsic Si is grown epitaxially (i.e., in order). The most common substrate is Si of the P type, which has a resistivity of 0.01–0.5 $\Omega\cdot$m (corresponding to acceptor densities of 10^{21}–10^{22} m^{-3}). Doping is generally produced during crystal growth.

4. *Slice preparation.* The desired plane is oriented ({100} or {111}) by X-ray diffraction and then cut into slices of 0.5–0.7 mm. After cutting, the surface is rough and must be mechanically polished with a mixture of Al_2O_3 powder and glycerin until one attains a uniformity better than 2 μm. This produces many defects. To remove them the previously oxidized surface is etched with HF.

5. *Epitaxial growth.* The few defects in the substrate do not allow its electronic use, but they do improve the epitaxial growth of a surface without defects. Of the many types, the most important are the following.

Chemical vapor deposition (CVD). Deposition of Si in the reaction $SiCl_4 + 2H_2 \rightleftharpoons Si +4HCl$. The most important issue is that the deposition must be done slowly.

Molecular beam epitaxy (MBE). Atoms of Si are evaporated by means of energetic electrons in an ultrahigh-vacuum (UHV) chamber so they impinge into the sample and are deposited. At the same time, the evaporation of dopants can create a semiconductor extrinsic layer. Again, the evaporation rate is highly significant.

The stages that follow are specific for each type of component needed. One must take care to mask zones that we do not want to modify with SiO_2 (stable and with little diffusion). For instance, in creating doped zones with other methods than those used for the substrate, we often make them by gaseous diffusion in the solid. Also, photolithography can be used (generally to make resistors), ion implantation (for dopants), metalization (to make electrical connections), and so on.

PROBLEMS

4.1. Copper density is 8.93×10^3 kg/m^3. Calculate the number of free electrons per cubic meter and deduce their average velocity if an electrical current of 1 A/cm^2 flows through the sample.

4.2. Silver density at $T = 300$ K is 10.5 g/cm^3. Calculate the average velocity of free electrons if an electrical current of 0.1 A/cm^2 flows through the sample.

4.3. Calculate R_H from the Lorentz force.

4.4. What is the meaning of the Hall effect in an intrinsic semiconductor? Why?

4.5. Calculate the electrical conductivity of an intrinsic semiconductor that has two types of electrons (negative charge) and one type of hole (positive charge) and whose conductivities associated with electrons are 10^{-4} and 2×10^{-4} $\Omega^{-1} \cdot$ m^{-1}, respectively; the conductivity associated with the holes is 2×10^{-4} $\Omega^{-1} \cdot$ m^{-1}.

4.6. In an extrinsic semiconductor the conductivity behaves differently at high, medium, and low temperatures. Explain the differences and interpret them.

4.7. Order the following materials from small to large electronic electrical conductivity at room temperature: intrinsic Ge, $Ni_{0.33}Zr_{0.67}$ (vitreous alloy), Cu, NaCl, N-type Ge.

4.8. Explain the main differences between unipolar and bipolar devices.

4.9. Why is the depletion zone electrically charged?

4.10. What is the main application of a P-N junction? Why?

4.11. Explain the microscopic mechanism of the Seebeck effect.

4.12. What limits the utility of the conductivity expression in the Drude model? Why?

Chapter Five

Mechanical and Thermal Properties

5.1 MECHANICAL PROPERTIES

To investigate the mechanical properties of a material one should apply to it different kinds of mechanical stimuli (usually mechanical forces). Examples of mechanical properties are strength, hardness, ductility, and stiffness. Frequently, mechanical properties are difficult to understand from microscopic theories, such as band theory, and they greatly depend on impurities and imperfections in the sample.[1]

We should first specify the most relevant concepts (stress σ and strain ϵ) and their relations in a sample. The *strain* produced in a sample is the mechanical response to the applied mechanical force per area unit, which is its *stress*. In materials engineering, the stress refers only to macroscopic procedures that deform a sample. Here, we use the most general meaning of this term to include internal stresses.

Stresses can be classified as tensile or compression stresses, where the stress is equivalent to a pressure (either positive or negative), or as shear stresses, where two forces are applied in the same direction but in different ways and at different points of application (i.e., a torque), or as torsion stresses, where two torques are applied in the same direction but in different ways and at different points of application. Each stress-strain curve is different from others due to the material's anisotropy and because the local stress distribution depends on the sample. Also, the application of stresses is different, and stress-strain curves also depend qualitatively on the kind of material, which makes a generic pictorialization difficult. Figure 5.1 is an idealized illustration.

To interpret stress-strain curves from the material structure and composition one must transform the measurements of stresses and strains to magnitudes proportional to the local and instantaneous stress and strain. For example, when we do a tensile test, the area where the stress is applied depends on the sample position and state, which in turn depends on time.

A material has *elastic* behavior in the regime—if it exists—where the stress is proportional to the strain. This proportionality is the *Hooke law*. The local proportionality constant should show the material's anisotropy. In this way the stresses may not have the same direction as the strains, and the proportionality constant may depend on the crystal direction. At a global level we could have different constants depending on the kind of stress. The constant corresponding to a tensile test is called the *stretch modulus* or *Young's modulus* E; the one corresponding to a shear test is called the *shear modulus* G. The ratio between lateral and axial strains is called the *Poisson coefficient* v. In isotropic materials $E = 2G(1 + v)$.

Elastic strain has its microscopic origin in the forces that appear in the lattice sites when they are moved from their equilibrium positions. Thus, the stretch modulus is proportional to the force (stress) produced when an atom or ion is displaced a fixed distance (strain). Then, in materials with strong bonds (e.g., ceramics) the moduli are higher than in materials

---

[1]For more detail see chapter 3.

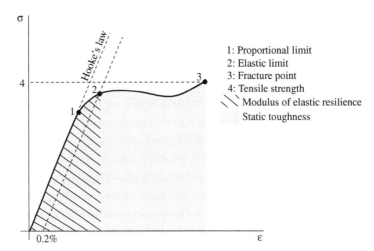

1: Proportional limit
2: Elastic limit
3: Fracture point
4: Tensile strength
Modulus of elastic resilience
Static toughness

Figure 5.1 Sketch of an ideal generic stress-strain curve. Important parameters are pointed out.

with weaker bonds (e.g., metallic materials or even more polymers). The moduli define a property called *stiffness*.

Notice that an elastic material goes back to its original state when the stresses are released, and the stress to keep a fixed strain is time independent. When a material manifests slight variations of this behavior, it is called *inelasticity*. When deviations from elastic behavior are large the material is viscoelastic (some polymeric materials are viscoelastic) owing to its intermediate behavior between an elastic material and a viscous fluid. When strains are not elastic they can show hysteresis, where the sample's properties depend on its history.

In general, one states that a material has *plastic* behavior if the applied stress is not proportional to the strain, and thus part of the strain is permanent. The mechanism of plastic strain is activated when interatomic bonds restructure (i.e., disappear, appear, change, etc.) and microfractures and microjoints are created.

The plastic behavior defines the *fluency*: the property where the sample, or a part of it, flows. What is more, the *proportional limit* is defined as the stress where the material is no longer elastic. This point is also known as the *elastic limit*. Some investigators conventionally define the latter as the stress at which the stress-strain curve intersects a parallel to the line determining the elastic behavior that passes through the point of zero stress and a fixed strain. This fixed strain is usually 0.2% (the difference between this elastic limit and the proportional limit is invaluable in materials engineering and indicates the plastic deformation resistance).

The *tensile strength* is the stress at the maximum of the stress-strain curve. It corresponds to the maximum tensile stress exerted on a sample without *fracture* (or breaking) of the sample. The fracture of a material may be *brittle* (produced before appreciable plastic strain appears) or *ductile* (produced after appreciable plastic strain). Ductility is the property that indicates whether a material can be plastically deformed without breaking. The fracture mechanism is complex and includes phenomena of mechanical, thermal, sound, electronic, and other types; it strongly depends on material imperfections (chapter 3).

The *resilience* is the capacity of a material to absorb elastic energy when it is strained and to release it when the stress is released. The resilience modulus U_r is a measure of this capacity. Resilient materials have a high elastic limit and a low stretch modulus.

The *toughness* is the capacity of a material to absorb energy, elastic or not, before the fracture point. A tough material has high ductility and strength. We can distinguish experimentally between static and dynamic toughness. The latter represents either the impact strength or the strength when a notch exists. The *hardness* is a measure of the strength in local plastic deformation. Several scales measure hardness; the conversion between them depends on the material and the measurement conditions.

5.2 PHONONS

Ions in a crystal lattice have an interaction potential corresponding to the interatomic forces. This potential can be approximated as the sum of two inverse powers of the interatomic distance. In this instance, two free parameters must be adjusted: $V = -C_1/r^n + C_2/r^m$ with $m > n$ (figure 5.2). These potentials have minima at the equilibrium atomic distance at 0 K. Near the minimum the potential is similar to a parabola, which corresponds to the potential of a spring, as given by Hooke's law. Consequently, for the small vibrations coming from the thermal agitation of a solid phase, we discover that the solid's behavior with respect to the lattice sites—if the sites do not change their positions—may be compared to a system of particles at the ends of interlocking springs (figure 5.3).

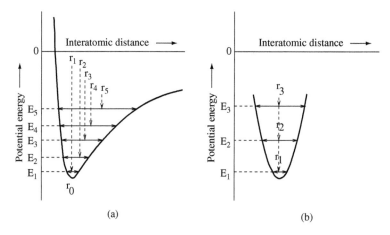

Figure 5.2 (a) Potential energy as a function of interatomic distance. The increase of the mean interatomic distance is shown as the temperature goes up. (b) Parabolic approximation of the interatomic potential.

The interlocking springs mean that the thermal agitation at different lattice sites is coupled. In this way, thermal agitation is indicated by normal modes in the lattice (i.e., stationary waves) that indicate the lattice sites' collective vibration. These normal modes are discrete, partially due to boundary conditions. Even if the solid is infinite, the modes are discrete because their energy is quantized. The energy quantum for vibrations is the *phonon*.

The dispersion relation of the phonon modes in monatomic periodic lattices has only one branch; in lattices with more than one chemical species the relation is split in various ways. The first branch, with lowest frequencies, is analogous to the acoustic branch and the associated acoustic modes. The following branches have higher frequencies than do acoustic modes: optical branches and associated optical modes.

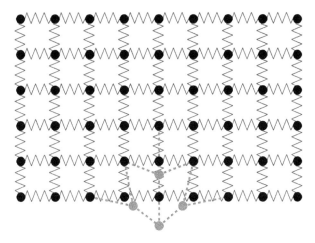

Figure 5.3 Sketch of lattice sites joined by springs. When a site is moved the sites surrounding it move too.

5.3 THERMAL PROPERTIES

Thermal Capacity and Specific Heat

The *thermal capacity* shows the heat storage capacity of a sample. It is represented as the energy required in a specific process (e.g., at a constant volume C_v, at a constant pressure $C_p > C_v$, etc.), to increase the sample temperature by 1 degree. Usually, the intensive magnitude is used: the *specific heat* c_v.

A solid's main mechanism for absorbing thermal energy is the lattice vibration in a crystal or the atomic or ionic vibration around their equilibrium positions. Thus, at $T = 0$ K the atoms or ions are at rest in the lattice sites. When the temperature is increased (i.e., when the system receives thermal energy) the atoms and ions begin to vibrate in a harmonic way around the equilibrium positions. (It is reasonable to suppose that the potential is parabolic near the minimum.) The higher the temperature, the more vibrations or thermal agitation we find. Since the atoms or ions interact, their vibrations are not independent; a lattice vibration is like an elastic wave (when quantized, a phonon). Such phonons indicate how large the elastic wave energy is and how many waves we have.

The phonon contribution to the specific heat consists of behavior at low temperatures of the following form:

$$c_v = AT^3,$$

A being a constant that does not depend on the temperature (problem 5.5). The Debye temperature θ_D is the temperature that divides the regime described above from the regime where the (molar) specific heat saturates at a value of

$$c_v = 3N_A k_B$$

which is the Dulong-Petit law, N_A being Avogadro's number. Most materials have a Debye temperature lower than room temperature; thus the phonon heat is constant. The thermal energy $k_B \theta_D$ is the maximum quantum of energy that can excite a lattice vibration, although in the lattice it is impossible for waves with small wavelengths to propagate.

The *electronic specific heat* is the contribution of Fermi gas heat. It can be obtained through purely quantum statistics. It is (at $T \sim T_{\text{room}}$)

$$c_v = \frac{\pi^2}{3} k_B^2 T g(E_F),$$

where g is a function of only the Fermi energy and represents the density of states per unit volume; for free electrons it is $g(E_F) = 3n/2E_F$. This heat is much lower than those of the other contributions, except in metallic materials whose temperatures are near 0 K.

Other contributions to the specific heat are magnetic due to the absorption or release of thermal energy by magnetic dipoles when they become ordered or disordered. This phenomenon and its applications are taken up in sections 6.4 and 7.1.

It is worthwhile noting that the thermal capacity is the key property in a sample. The reason is that the specific heat is an intensive magnitude. It does not indicate a sample's size, nor the true thermal capacity.

Thermal Conductivity

The *thermal conductivity* κ is the capacity of a material to transport heat, except by radiation or convection (where there is mass transport as well).

Heat conduction in solids has two mechanisms: the first is due to phonons, and the most important for metallic materials is due to the gas's thermal agitation. This takes place because electrons are scattered less and can reach higher velocities. In this sense the thermal conductivity in metallic materials is roughly proportional to the electrical conductivity (the Wiedemann-Franz law [section 4.1]).

$$\kappa = \frac{\pi^2}{3} \left(\frac{k_B}{e} \right)^2 T \sigma.$$

To calculate the electronic thermal conductivity, you should consider that it is related to the electronic specific heat—which is smaller than usual—and also to the electrons' mean velocity, which is higher than in an ideal gas at room temperature. The two conditions cancel out the errors in a classical calculation and yield an approximate value. The conductivity in metallic materials lies between 20 and 400 W/mK.

In dielectric materials the thermal conductivity is due only to phonons because an electron gas does not exist. Since phonon scattering by lattice imperfections is efficient, the conductive heat transport is much lower than in metallic materials, between 2 and 50 W/mK. As phonon scattering increases when the temperature rises, the conductivity declines at higher temperatures, until any nonconductive mechanism (infrared radiation) enhances heat transport. If the material is porous, the conductivity is even smaller, such as 0.02 W/mK, because gas convection inside the pores is small. The most common nonmetallic material and best heat conductor is BeO, where phonon scattering is small.

Thermal Expansion

The *thermal expansion* indicates how materials expand or contract if the temperature is raised or lowered. Thus, a material's length varies when its temperature rises by 1 degree. This is called the linear expansion coefficient α_l, which in anisotropic materials depends on direction. The variation in a material's volume when its temperature rises by 1 degree is called the volume expansion coefficient α_v. If the material is isotropic, then $\alpha_v = 3\alpha_l$. Most expansion coefficients are positive: an exception is water between 0 and 4°C.

Thermal expansion can be understood by examining the interaction potential, which can be compared to a parabola near its minimum. But if we go farther it loses symmetry and consequently its harmonic character (figure 5.2); this means that the interatomic distance expands if the temperature also rises.

Notice that the thermal expansion coefficient is highly significant in many applications. It is necessary to have small coefficients in materials that should have a determined size, or in materials exposed to thermal shocks, which are thermal stresses due to fast or inhomogeneous changes of temperature. In these circumstances, isotropic and small coefficients are of interest. For joints between materials we need similar coefficients to avoid thermal stresses. There are instances, as with earthenware steel exposed to strong thermal stresses, when the match of both coefficients is in a broad range of temperatures. Joints of glass-metal are used as electric bolts once known as Kovar$^{®}$. Nowadays we obtain ceramics by mixing materials that have positive and negative expansion coefficients. These allow us to obtain an expansion coefficient of almost zero.

Common conductors have linear expansion coefficients between 5×10^{-6} and $25 \times 10^{-6} \, °C^{-1}$. Those of ceramics are an order of magnitude smaller, but those of polymers are an order of magnitude higher than those of conductors.

PROBLEMS

5.1. Discuss this: All ductile materials are soft.

5.2. Discuss this: All soft materials are ductile.

5.3. Distinguish a ductile fracture from a brittle one.

5.4. What mechanical properties depend on toughness?

5.5. The (molar) vibratory, or phonon, specific heat at low temperatures (much smaller than the Debye temperature θ_D) is $c_v = AT^3$, where $A = 12\pi^4 N_A k_B / 5\theta_D^3$ is N_A, the Avogadro number. Calculate the Debye temperature for copper, whose specific heat at 15 K is 4.60 J/kg · K. Discuss the coherence of the solution.

5.6. Why is it possible to use two metallic strips of materials glued along their length to serve as part of a thermostat?

5.7. Why does a metallic material feel colder than a nonmetallic material, even when the materials are at the same temperature, if this temperature is lower than that of the human body?

5.8. The Dulong-Petit law reflects the number of vibratory degrees of freedom in a solid. Discuss this fact.

5.9. Using the arguments of the preceding problem, discuss why the Dulong-Petit law is not verified at low temperatures.

5.10. Order the following materials from low to high thermal conductivity: BeO, Al_2O_3 (high porosity), Ag, Al_2O_3 (low porosity).

5.11. Explain how the volume expansion coefficient and the linear one are related in an isotropic solid. What would the relationship be for an anisotropic solid?

Chapter Six

Magnetic Materials and Dielectrics

Electromagnetism is the part of physics that has to do with the electrical and magnetic interactions in a vacuum and inside matter. From the classical point of view, these interactions are modeled mathematically by the Maxwell equations. If it is necessary to include quantum effects in the system, then quantum electrodynamics should be used—but we do not use it here.

The Maxwell equations within matter can be reduced to the Maxwell equations in vacuum if we have an electrical field created by atomic electrical charges (which lead to polarization) and if we have a magnetic flux density owing to internal currents and intrinsic magnetic momenta (related to the spins).

In this way (in the MKSA unit system[1]), we get

$$\vec{B} = \mu_0(\vec{H} + \vec{M}),$$

$$\vec{D} = \epsilon_0\vec{E} + \vec{P},$$

where the *polarization* $\vec{P}$ and the *magnetization* $\vec{M}$ contribute to the equations of the internal electrical and magnetic fields. The Maxwell equations indicate that

- There are charges (monopoles) from electrical fields:

$$\nabla \cdot \vec{D} = \rho.$$

- There are no charges (monopoles) from the magnetic field:

$$\nabla \cdot \vec{B} = 0.$$

- A changing magnetic field creates an electrical field:

$$\nabla \times \vec{E} = -\frac{\partial \vec{B}}{\partial t}.$$

- Electrical currents and changing electrical fields create a magnetic field:

$$\nabla \times \vec{H} = \vec{J} + \frac{\partial \vec{D}}{\partial t}.$$

The electromagnetic field acts on a charge q and exerts the *Lorentz force*:

$$\vec{F} = q(\vec{E} + \vec{v} \times \vec{B}).$$

All the magnetism and electricity inside matter can be described by the polarization $\vec{P}$ and the intensity of magnetization $\vec{M}$. On the whole, if the system is isotropic, $\vec{P}$ and $\vec{M}$ verify that

$$\vec{D} = \epsilon\vec{E} = \epsilon_0\epsilon_r\vec{E},$$

---

[1] In this chapter we use this unit system.

$$\vec{H} = \frac{1}{\mu}\vec{B} = \frac{1}{\mu_0\mu_r}\vec{B},$$

where ϵ is the permittivity and μ is the permeability.[2] In anisotropic media these are tensors[3] and the relative permittivity ϵ_r and permeability μ_r may depend on $\vec{E}$ and $\vec{B}$; their value is 1 in vacuum.

Other parameters can be defined, in particular, the susceptibility χ and the electrical susceptibility χ_e, which indicate how the polarizations change with respect to the fields. The result? Different properties of magnetic or electrical materials can be related to the behaviors of their susceptibilities:

$$\vec{P} = \epsilon_0\chi_e\vec{E},$$

$$\vec{M} = \chi\vec{H}.$$

As is well known, there are no magnetic monopoles. The sources of the magnetic field are electrical currents and intrinsic magnetic dipoles. Inside matter, these sources are, first, currents generated by electron movement and thus connected to angular momentum and, second, the intrinsic magnetic moments of the electrons and the atomic nucleus (and hence their spins). An electron's magnetic moment caused by its spin is the Bohr magneton $\mu_B = e\hbar/2m_e$.

Magnetic materials belong to three categories: *paramagnets*, *diamagnets*, and *ferromagnets*. Paramagnets and diamagnets do not show any magnetization in zero field. The interaction between magnetic dipoles gives ferromagnetic materials, where it is possible to obtain a nonzero magnetization without an applied field.

Paramagnets and ferromagnets are made of permanent microscopic magnetic dipoles. But diamagnets are a response to external fields in materials having molecules with a zero magnetic dipolar moment. The magnetization is the sum of a region's magnetic dipole moments.

6.1 DIAMAGNETISM AND PARAMAGNETISM

Paramagnetic materials have a small positive magnetic susceptibility χ. This stems from the tiny interaction between material magnetic dipoles. The interaction is not strong enough to align the magnetic moments and lead to finite magnetization because of the effects of thermally induced disorder, which results in a zero mean magnetic moment. Examples of paramagnetic materials are β-Sn, W, Al, Pt, and Mn.

When an external magnetic field is present, the magnetic dipoles try to align with it in the same way that at 0 K all dipoles are oriented; the resulting magnetization is called the saturation magnetization M_s. But at $T > 0$ K there is a disordering effect, which means that only a fraction of dipoles stay oriented.

In this instance the magnetization is

$$M = \frac{1}{V}\frac{\sum_i \vec{\mu}_i \cdot \vec{B}}{B},$$

where $\vec{B}$ is the magnetic field and $\vec{\mu}_i$ is the magnetic moment of each particle. Then,

$$\vec{\mu}_i \cdot \vec{B} = \mu\cos(\alpha_i)B$$

---

[2]The Greek symbol for permeability μ is confusing with respect to the magnetic dipole moment. Therefore, we will express permeability as the permeability in a vacuum μ_0 times the relative permeability μ_r.

[3]In this instance, we mean that they can be viewed as matrices.

and thus

$$M = n\mu \langle \cos(\alpha_i) \rangle,$$

where n is the number of magnetic dipoles per unit volume.

The *magnetic dipolar energy* is $E_i = -\mu B \cos(\alpha_i)$. It is related to the distribution of particles in relation to their specific temperature. So, if we consider an equilibrium Boltzmann distribution:

$$\langle \cos(\alpha_i) \rangle = \frac{\sum_i \cos(\alpha_i) e^{-E_i/k_B T}}{\sum_i e^{-E_i/k_B T}} = \frac{\int_{-1}^{1} x e^{ax} dx}{\int_{-1}^{1} e^{ax} dx},$$

where $a = \mu B/k_B T$, then $\langle \cos(\alpha_i) \rangle = L(a)$, where $L(x) = \coth(x) - 1/x$ is the Langevin function.

Thus, $\chi = M/H = \mu_0 M/B = \mu_0 n(\mu/B)L(\mu B/k_B T)$. For temperatures approaching 0 K and nonzero fields, the Langevin function approaches 1, and the magnetization approaches the saturation value $n\mu$. For high temperatures the Langevin function approaches zero and there is no magnetization. The magnetization curve is represented in figure 6.1.

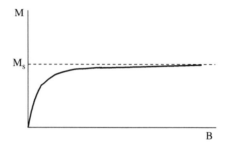

Figure 6.1 The magnetization curve for a classical paramagnet.

For small values of $\mu B/k_B T$, one gets the *Curie law*, which states that the susceptibility does not depend on the applied field:

$$\chi = \mu_0 n \frac{\mu^2}{2k_B T}.$$

If the magnetic moments are quantized, only discrete energies are allowed. In this instance, the magnetization is slightly different and is characterized by Brillouin functions instead of the Langevin function.

Added to this paramagnetic behavior is that corresponding to the magnetic moment of the electron gas in metallic materials. Here, it is necessary to consider the band quantum model and quantum statistics (thus, we must apply the Pauli exclusion principle to the electrons). The electrons have two states of spin ($1/2$ and $-1/2$). This means that electron gas energies in bands are displaced in one direction for electrons with a spin parallel to the field and in the other direction for electrons with a spin antiparallel to the field. At first, $N_\uparrow = N_\downarrow$. Yet the electrons progress toward thermal equilibrium in such a way that they reach a maximum energy equal to the Fermi energy, where they finally arrive at equilibrium with $M = (n_\uparrow - n_\downarrow)\mu_B$, $n_\uparrow$ and $n_\downarrow$ being the number of electrons (per unit volume) with energies from $-\mu_B B$ to E_F and from $\mu_B B$ to E_F, respectively (figure 6.2). Knowing that $\mu_B B \ll E_F$ (for realistic fields), we figure that $n_\uparrow \approx n_\downarrow$ can be reached, and thus the magnetic susceptibility is much smaller than it is for nonmetallic materials.

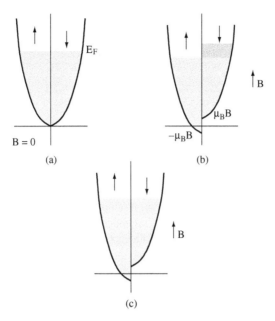

Figure 6.2 Sketch of the origin of metallic paramagnetism. (a) Initial energy bands without fields. (b) Displaced energy bands due to magnetic fields; unstable occupation of states. (c) Final steady situation with an applied magnetic field.

All kinds of materials present diamagnetic behavior. Because this effect is much smaller than the paramagnetism in materials with permanent magnetic dipoles, diamagnetism is relevant only for materials without permanent magnetic dipoles; in other cases diamagnetism is screened by the main effect.

Purely diamagnetic materials are those that present a small negative magnetic susceptibility. The material is repelled by magnets because they induce a magnetic moment opposite to the applied field.

Diamagnetism can be understood with the help of the Lenz law. If an atom has the same number of electrons moving in both directions, and if there is an equal number of electrons with spin up and down, then the global permanent magnetic moment is zero. This will happen in molecules with closed shells of electrons. It leads to a zero net angular momentum, as occurs in Bi, Be, Ag, Au, Ge, Cu, Si, MgO, and so forth. If we apply a magnetic field, the internal currents change and the changes are opposite to the variations of the magnetic flux. This creates a net field—and hence we measure a net magnetic moment—opposite to the applied one. As a consequence, the material repels magnets. All these behaviors occur in both directions of movement without canceling themselves, and so magnetic dipoles are induced.

If the material is paramagnetic, the field's first effect is to orient the permanent magnetic dipoles in the field direction. After that, the diamagnetic component appears, but is much smaller than the paramagnetic one (of the order of μ_B). Hence, diamagnetism is relevant only at high temperatures where paramagnetism responds to the disordering effect of thermal vibration.

A superconductor, as we see in chapter 7, is a perfect diamagnet. Its magnetic susceptibility is, then, $\chi = -1$.

To guess the diamagnetic magnetization it is necessary to realize that, when a magnetic field B is applied, the force changes in such a way that it equals the centripetal force $evB = dF = 2mv\,dv/r$. Therefore, $dv = erB/2m_e$. The magnetic moment of a moving charge is $\mu = (1/2)evr$; consequently, the change of magnetic moment will be

$$\Delta\mu = \frac{1}{2}er\,dv = \frac{e^2r^2B}{4m_e} \sim 10^{-5}\mu_B,$$

as we predicted.

6.2 FERROMAGNETISM, FERRIMAGNETISM, AND ANTIFERROMAGNETISM

Ferromagnetic materials (Fe, Co, Ni, their alloys, Gd, Dy, etc.) have large positive values of the magnetic susceptibility that depend on the magnetic field. These materials behave this way because of interactions among spins (called exchange), and they may have net nonzero $\vec{M}$ at $\vec{H} = 0$. Also, they have a complex behavior in comparison with paramagnets. Thus, by adding this interaction to the magnetic ingredients of paramagnets, we can obtain the materials discussed here.

In a ferromagnetic material a spin is usually aligned in a particular direction (that of the applied field if there is one, or randomly if there is not). This dipole interacts enough to align the neighboring spins in the same direction and form *magnetic domains* where the net magnetization is nonzero.

Theories that describe ferromagnetic materials are much more complex than those used for paramagnets. Yet ferromagnetism can be understood qualitatively by considering that the magnetic field created by the magnetization is partially derived from the magnetization itself. Thus, the Weiss molecular field theory proposes that

$$M = n\mu L\left(\frac{\mu B_{\text{eff}}}{k_B T}\right) = n\mu L\left[\frac{\mu(B + \lambda M)}{k_B T}\right]$$

where λ is the molecular field coefficient.

Notice that if $B = 0$ we obtain $M = n\mu L\,[\mu\lambda M/k_B T]$, an equation that has distinct behaviors at different temperatures (figure 6.3).

Therefore, T_c, called the Curie temperature, is defined in such a way that for $T < T_c$ the material presents a net magnetization, even without an applied field (figure 6.4). Consequently, the material behaves ferromagnetically. For $T > T_c$ the material behaves paramagnetically according to the Curie-Weiss law:

$$\chi = \frac{C}{T - T_c}$$

where C is the Curie-Weiss constant. The Curie temperature T_c is 770 K for Fe, 358 K for Ni, 1131 K for Co, and 293 K for Gd.

Ferromagnets show hysteresis, which reflects the dependence of the magnetic phenomena on their history. As an example, a hysteresis loop is presented in figure 6.5, where B_r is the *remanent flux density* (i.e., the remaining field when there is no applied H), and H_c is the *coercive force* (which is the applied field needed to obtain a field or magnetization equal to zero).

The three parameters B_r, B_s, and H_c are the ones that determine most applications, as well as the area inside the hysteresis loop, which is the energy density dissipated in the loop:

$$E_{\text{diss}} = \oint \vec{B}\cdot d\vec{H}.$$

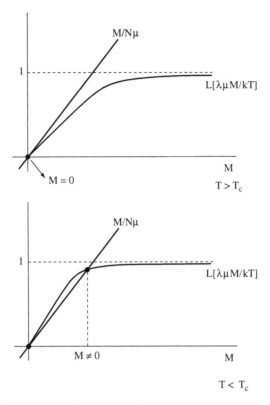

Figure 6.3 A graphic solution to the magnetization equation (at zero applied field) corresponding to the classical Weiss molecular field.

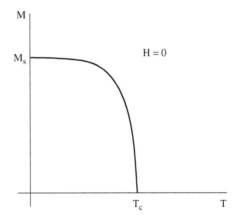

Figure 6.4 Magnetization versus temperature at zero field. The critical temperature is shown.

A ferromagnet is composed of magnetic domains. All of them are at saturation flux density or magnetization. The magnetic properties of these materials can be understood from the existence and evolution of the magnetic domains and, specifically, from the *anisotropy constant* (which tells us how anisotropic the sample is) and the *saturation magnetostriction constant* (which indicates the sensitivity of the magnetization to internal stresses in

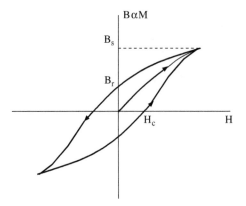

Figure 6.5 Hysteresis loop for a ferromagnet.

the sample) (problems 6.9 and 6.10). The existence of domains reveals that sometimes ferromagnetic samples show no spontaneous magnetization.

There are two kinds of ferromagnetic materials: soft magnetic materials with $H_c < 10\,\mathrm{Oe}$ and narrow hysteresis loops, such as FeSi, NiFe, CoFe, transition metal (Fe, Co, Ni)–semimetal (B, C, Si) alloys, and soft ferrites (amorphous or crystalline), and, on the other hand, hard ferromagnetic materials with $H_c > 10\,\mathrm{Oe}$ and wide hysteresis loops, such as transition metal–rare earth (Gd, Tb, Sm) alloys of large anisotropy, and hard ferrites. If in the classical treatment of ferromagnetism the band theory is included, the predicted outcomes are more numerous.

An important concept is the demagnetizing field caused by a shape anisotropy. Inside a magnetized ferromagnet there is a field $\vec{H} = \vec{H}_a - N\vec{M}$, where $N \in (0, 1]$, a constant that depends on the sample's direction and shape (figure 6.6).

This means that a magnet is demagnetized spontaneously except when the magnetic circuit is closed with a ferromagnetic material (figure 6.7).

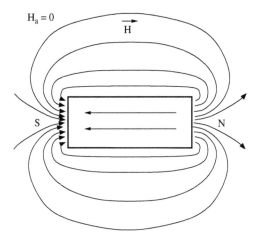

Figure 6.6 A demagnetizing field caused by a shape anisotropy.

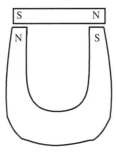

Figure 6.7 A short-circuit of a magnetic field with a ferromagnet. Magnetic poles are induced in such a way that the demagnetizing field abruptly decreases.

Antiferromagnetic materials (e.g., MnO, FeO, CoO, NiO, Cr, etc.) have interacting magnetic dipoles that try to align in an antiparallel position, so that even below the critical temperature (the Néel temperature) $M(H = 0) = 0$. The Néel temperature defines the order-disorder transition in the subnets of equally oriented magnetic dipoles. The transition is not directly measurable from the magnetization values; it is examined instead with nuclear magnetic resonance techniques.

Ferrimagnetic materials have multiple components of antiferromagnetic type, but with the magnetic dipoles having different magnitudes in one direction from the other; they yield a net magnetization (figure 6.8). Materials like γFe_2O_3 show ferrimagnetic behavior that is quite useful in magnetic recording; other examples are $5Fe_2O_3 \cdot 3R_2O_3$ where R is a rare earth like Y, and $MO \cdot Fe_2O_3$ where M is a transition metal, and so on.

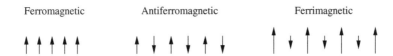

Figure 6.8 Sketch of different kinds of magnetic materials with strong interactions.

Ferromagnetic-like materials can create frustration, where the magnetic dipole system cannot reach the minimum energy state (figure 6.9). There are other magnetic properties in disordered systems like ferrofluids (a stable colloidal dispersion of magnetic monodomains), spin glasses, and so forth. We do not consider such properties here. They lead to other magnetic materials like superparamagnets.

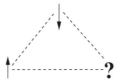

Figure 6.9 Frustration in an antiferromagnetic system located in a triangular lattice: two spins are imposed and the third is undetermined.

6.3 MAGNETIC RESONANCES

All magnetic resonances are based on resonant excitation through an oscillating magnetic field (or electromagnetic radiation, or both) of magnetic dipoles, which make transitions between energy levels ΔE in a magnetic system. Therefore, $\Delta E = h\nu$. The idea is to vary $\vec{H}$ with increasing frequency ν until the transition is produced. It is then that we know the relative positions and magnitudes of the energy levels. There are different magnetic resonances,[4] among them the following

1. Electronic paramagnetic resonance (EPR). This resonance works in the microwave range (10^9–10^{11} Hz). With it we study magnetic dipoles of electronic origin. This resonance is useful for identifying molecules, atoms (with $\vec{J} = \vec{L} + \vec{S} \neq 0$), ions, nonstoichiometric imperfections in solids and in extrinsic semiconductors, and so on. The resonant frequencies depend on the atoms' local surroundings. The technique has a 1 MHz resolution and measurement times greater than 10^{-9} s.

With EPR measurements one can determine spectroscopic parameters (energy levels, crystal field parameters) to determine spin, isotopic abundance, local symmetries, spin-lattice interactions, magnetic interactions, nonequilibrium phase transitions, and biochemical processes in photosynthesis, and to make quality control assays for irradiation. Another application of electronic paramagnetic resonance is for making microwave ovens.

2. Nuclear magnetic resonance (NMR). This resonance works in the radio-frequency range (10^6–10^9 Hz) and enables investigators to study nuclear magnetic dipoles. It is useful in nuclei like ^{2}H, ^{19}F, ^{1}H, ^{13}C, ^{31}P, ^{14}N, ^{15}N, and ^{33}S. The frequencies depend on the screening of the crystal field and of the electrons to the nucleus, so information is obtained about the chemical environment. The technique has a 1 kHz resolution and its measurement times are lower than 10^{-11} s. Its main application is in medicine, although one can observe the different concentrations of different nuclei without affecting the material's biochemical properties. This does not happen when using EPR.

3. Nuclear resonance γ (also called Mössbauer spectroscopy). This kind of resonance is based on the absorption or emission, or both, of γ radiation by nuclei of different materials in such a way that the relevance of the crystal field can be measured (due to the hyperfine interactions).

As absorption occurs only if the transition energy is the same as the energy of the incoming radiation, the sample must be moved to increase (or decrease) the system's kinetic energy. If we know how fast the system moves, we can deduce the influence of the crystal field with precision. This is possible so long as the sample temperatures are low enough to avoid movement of the nuclei due to thermal agitation. The resolution should be $|E - E'| \geq \Gamma \approx 10^{-8}$ eV.

4. Other techniques include emission Mössbauer and conversion electron Mössbauer spectroscopy.

6.4 APPLICATIONS

Some applications are the following.

Measurement equipment. We have available a lot of measurement tools; among them are EPR, NMR, Mössbauer spectroscopy, and other magnetic sensors, (using properties

---

[4]Magnetic resonance, as well as spectroscopic methods, is a technique used in several branches of physics, such as electromagnetism, physical optics, solid-state science, surface science, quantum mechanics, and so forth.

like magnetostriction, magnetotorsion, magnetoresistance, the Hall effect (appearance of a voltage perpendicular to $\vec{H}$ and to $\vec{I}$), and so on).

Magnets. H_c, M_r, and T_c should be high. They are used to create an *action at a distance* or to supply a magnetic flux. Their applications include acoustic transductors, motors and electric generators, stepper motors, magnetomechanics (material separators, magnetic material transportation), focusing systems of charged particles, and so on. They should be comprised of materials with a high anisotropy constant, that is, hard magnetic materials such as alloys of rare earth elements and transition metals.

Magnetic recording. To make a magnetic recording we need to use small particles with attainable switching fields (nowadays the size is of the order of 1 μm, which is expected to climb to 50 Å by 2020). Classical relaxation is due to the temperature and quantum relaxation is caused by the tunneling effect. To get durable materials H_c, M_r, and T_c should be sufficiently high.

Electrical transformers. Mutual induction between two electromagnets enables one to manufacture transformers. Small leakage at high frequencies is required; thus, the materials should have a narrow hysteresis loop, or be soft magnetic materials, with small anisotropy (e.g., the amorphous ferrites).

Magnetocaloric effects. These effects couple magnetic fields with heat fluxes. To be exact, one of these is adiabatic demagnetization. This is useful for paramagnetic materials at an initial temperature T_0 without an applied magnetic field. The whole process is as follows:

1. Application of the magnetic field in a thermal bath to keep a constant temperature. Here the sample energy has a thermal component and a magnetic one.

2. Insulation of the magnetized paramagnetic sample.

3. Adiabatic suppression of the magnetic field. This step is adiabatic demagnetization. When the field is eliminated the paramagnetic sample becomes disordered and its magnetic energy increases. Because the total energy remains constant due to isolation, the thermal energy has to decrease, which reduces the thermal vibrations of the lattice sites and consequently the sample temperature to a value $T_1 < T_0$.

This process is efficient at low temperatures. For instance, it allows a paramagnet to cool from 1 K to 1 mK. For temperatures appreciably higher, like room temperature, there are other more efficient methods.

Variable viscosity systems (ferrofluids). Here $\nu = \nu(\vec{H})$.

Magnetooptic recording (Kerr effect). This is based on modifying $\vec{M}$ by using electromagnetic radiation.

Amorphous magnetic materials.

6.5 DIELECTRICS

Dielectrics are the most relevant materials for examining electrical properties different from those of semiconductors and conductors. These materials can be defined as dielectric and as nonconductors, which means that the electronic charge is not free: The electrons are localized, like the ions. Because the charge is localized, dielectrics can accumulate or store it. Thus, ionic-covalent and van der Waals bonds between closed-shell atoms lead to dielectric materials, where the valence bands are full of localized electrons and the conduction bands are far enough from the valence bands to be empty, because electronic promotion is small. The few promoted electrons in a good dielectric permit the application of fields of roughly hundreds of volts per centimeter but generate currents as low as only 10^{-9} A. Dielectrics can also become charged by friction. Although this phenomenon is still not completely understood, a tentative explanation is that in materials like common glass the external surfaces cut through the lattice in such a way that the free bonds can lose electrons during a friction process, so that the glass acquires a positive net charge. Polymeric materials like resins, bakelite, silk, and nylon can lose H^+ ions and acquire a negative net charge. When their surface is charged, all these materials lose the charge easily because of the OH^- and H^+ ions in damp air.

Dielectric Properties

Permittivity ϵ ($\vec{D} = \epsilon \vec{E}$).

This quantity measures a material's ability to store a charge. Thus, the capacitor capacity is $C \equiv Q/V \propto \epsilon = \epsilon_0 \epsilon_r$. The permittivity is a complex quantity if the applied potential is variable: $\epsilon_r = \epsilon' - \iota \epsilon''$. We have $C \propto \epsilon'$ and $\tan(\delta) = \epsilon''/\epsilon'$, which is the loss factor of the dielectric. The smaller it is, the better the insulating material.

Polarizability α.

This is the electrical dipole moment per field unit and $\alpha = \epsilon_0 \chi_e/N$, where N is the number of dipoles per volume unit. It can be expressed as the sum of polarizabilities corresponding to different contributions. In this way, $\alpha = \alpha_e + \alpha_a + \alpha_d + \alpha_i$, where α_e is the electronic or optical polarizability due to the deformation of the electronic cloud, α_a is the atomic polarizability owing to the elongation of dipoles, α_d is the orientational polarizability created when dipoles rotate to align with the field direction, and α_i is the interface polarizability resulting from the accumulation of charge at planar defects or limit surfaces.

The permittivity clearly has a different behavior at different frequencies corresponding to the contributions mentioned above (figure 6.10). Thus, α_e relaxes at ultraviolet or optical frequencies, α_a at infrared frequencies, α_d at radio frequencies, and α_i at frequencies of the order of 1 Hz.

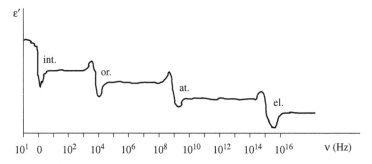

Figure 6.10 Permittivity spectrum of a hypothetical material. "int." indicates interface, "or." orientational, "at." atomic, and "el." electronic contributions to permittivity.

Dielectrics can be classified according to their polarizabilities:

Nonpolar materials, where $\alpha_a = \alpha_d = 0$. Without an electrical field there are no dipoles. All monatomic materials are nonpolar.

Polar materials, where $\alpha_d = 0$. Ionic solids, like alkali halides, are polar materials.

Dipolar materials. Water and hydroxyl and carbonyl groups are dipolar.

Piezoelectricity.

Materials having this property (materials with less symmetry: 20 of the 32 crystal point groups) present a mechanical stress in response to an electrical field, and vice versa. The piezoelectric coefficient can be defined as

$$d = \left(\frac{\partial P}{\partial T}\right)_E = \left(\frac{\partial S}{\partial E}\right)_T$$

and another coefficient is defined as

$$g = \left(-\frac{\partial E}{\partial T}\right)_P = \left(\frac{\partial S}{\partial P}\right)_T,$$

where T is the stress and S is the strain that satisfy $d = \epsilon g$.

At the practical level, the coupling coefficient k can be defined as the conversion energy efficiency, where $k \propto \sqrt{dg}$.

Quartz (SiO_2) is the best known of the piezoelectric materials (figure 6.11); in fact, by cutting quartz into slices one can build oscillators used in watches. Quartz is common, cheap, and not pyroelectric (and thus its properties do not depend much on temperature). Piezoelectric ceramics can have coefficients as high as 50 times those of quartz. They are useful in microphones, force meters (force transductors), spark generators, lighters, needles, and so on.

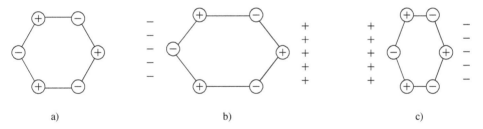

Figure 6.11 Mechanism of piezoelectricity in quartz. (a) Unit cell without deformation. (b) Stretched unit cell. (c) Squeezed unit cell.

Pyroelectricity.

This property is presented by 10 of the 20 piezoelectric symmetry groups due to their unique polar axis; they have spontaneous polarization. There is a one-to-one correspondence of an electrical field and a temperature change. The pyroelectric coefficient can be defined as

$$p = \frac{\partial D}{\partial T}.$$

In equilibrium, the depolarizing field that was created by the polarization is neutralized by the free charge through an external circuit. When the temperature becomes stabilized the current decays exponentially. Most pyroelectric materials are also ferroelectric (similar to ferromagnets in magnetism). Pyroelectric materials are generally useful for detecting temperature variations by infrared radiation owing to their high sensitivity (1 mK); hence they are used in alarms, infrared cameras, and the like.

Ferroelectricity.

This property consists of having spontaneous polarization at temperatures below the Curie critical temperature T_c. If the temperature is reduced below T_c at zero field, the polarized domains cancel out their total polarization. But if the sample is polarized above T_c, where a Curie-Weiss-like law is satisfied:

$$\chi_e = \frac{C}{T - T_c},$$

and afterward the temperature is lowered, a constant polarization field appears. At temperatures above the Curie temperature, ferroelectric materials transform to paraelectrics, whose behavior is similar to that of paramagnetic materials in magnetism. This analogy is also true for domains, for the depolarizing field coming from shape anisotropy, and for hysteresis, among other effects.

There are also antiferroelectric and ferrielectric materials.

There is no quantitative theory to explain ferroelectricity, but this phenomenon is qualitatively due to the interaction among electrical dipoles of the material. Because of the strong dependence of ferroelectricity on temperature, many pyroelectric materials are also ferroelectric, and vice versa.

PROBLEMS

6.1. Does diamagnetic susceptibility depend on temperature? How?

6.2. Does diamagnetic susceptibility depend on atomic number? How?

6.3. What role does the crystal field play in paramagnetism; that is, the anisotropy coming from the disposition of atoms in a solid rather than from being free in space?

6.4. Find T_c for the classical Weiss model.

6.5. Explain why adiabatic demagnetization is not used to cool paramagnetic materials at room temperature.

6.6. How can you demagnetize a ferromagnetic material?

6.7. The chemical formula of magnetite can be expressed as $(Fe^{2+}O)(Fe_2^{3+}O_3)$. We know that the Fe^{3+} ion leads to antiferromagnetic behavior and the Fe^{2+} ion to ferromagnetic, and that their dipole moments are, respectively, $5.9\mu_B$ and $5.4\mu_B$. Find the magnetization of magnetite at $T = 0$ K. ($\rho = 5180$ kg $\cdot$ m^{-3}, $P_{Fe} = 55.8$ g $\cdot$ mol^{-1}, $P_O = 16.0$ g $\cdot$ mol^{-1}.)

6.8. Explain the essentials of Mössbauer spectroscopy and its applications.

6.9. The energies involved in the behavior of a ferromagnetic material are the exchange interaction, which tries to align the magnetic dipoles, the magnetocrystalline anisotropy, which is due to a crystal anisotropy that makes some directions more sensitive than others to magnetic fields, the domain wall magnetic energy caused by the magnetic dipoles' shift in orientation across approximately 10^2 atomic layers, the magnetostatic energy, which is the magnetic field energy created by the sample, and the magnetostriction energy, which is the elastic energy due to the strains of magnetostrictions. Explain which situations minimize each of these energies separately.

6.10. Explain the origin of each energy in problem 6.9 related to the different anisotropies. Review the conditions necessary for these energies in order for a material to be magnetically soft or hard, an antiferromagnet, or not a ferromagnet.

6.11. Explain the piezoelectric effect in quartz.

6.12. Explain the possible microscopic mechanisms of pyroelectricity.

Chapter Seven

Superconductivity

7.1 INTRODUCTION AND APPLICATIONS

In 1911 Kamerlingh Onnes discovered that mercury resistance falls to zero at temperatures lower than 4.2 K. He called this property *superconductivity*. He also found that a high enough magnetic field destroys superconductivity. Later on, investigators observed that a persistent induced current was created in a ring formed by two different superconductors. Meissner and Ochsenfeld in 1933 showed that superconductors push out magnetic field lines, as we find below. In 1950, Fröhlich pointed out that phonon-mediated electron interactions are important in superconductivity. Only seven years later Bardeen, Cooper, and Schrieffer constructed the first model (known as BCS theory) that explains low critical temperature superconductivity. Finally, in 1986 high critical temperature (e.g., $T_c \geq 30$ K, $T_c \sim 125$ K) superconductors were discovered, but no theory currently explains high critical temperature superconductors.

Conductor state properties change abruptly for temperatures lower than the critical temperature. This proves the presence of a phase transition similar to the ones in ferromagnetic materials—the most ordered and stable state at low temperatures. The transition width is small (roughly 10^{-3} K) and strongly depends on crystalline imperfections and impurities.

Yet not every pure metal has superconductivity; or they do have it but in extreme conditions. Thus, W is a superconductor only for temperatures lower than 0.01 K. Si is a superconductor at temperatures lower than 7 K, but it must be under a pressure of 130 kbars. Materials in ferromagnetic states like Fe, Ni, and Co, are not superconductors. Also, good conductors like Cu or Ag are not superconductors. Fe in a nonmagnetic phase whose temperatures are lower than 2 K is a superconductor if we apply to it a pressure between 15 and 30 GPa: this zone corresponds to a bcc-hcp structural phase transition.

The properties characterizing a superconducting phase are several. In the first place the resistivity goes to practically zero values (10^{-29} $\Omega \cdot$ m). In an ideal conductor free electrons move like plane waves through the structure without losing momentum in the original direction. The resistivity in real conductors has two origins: atomic vibrations and imperfections. A perfect conductor at $T = 0$ K has $\rho = 0$. Superconductors have this property for $T \leq T_c > 0$ K; there is no dissipation and there is no voltage drop in the superconductor when a constant current passes through it. Note that, although the electrical resistance goes to zero, the impedance is greater than zero; that is, if a superconductor had time-dependent currents, a voltage between the superconductor terminals would appear. This happens because time-dependent currents rely on the inertia of Cooper pairs and normal electrons causing an electromagnetic field and hence a voltage. Nevertheless, this impedance is small in comparison with the impedance of normal conductors, but greater than that of the direct current in superconductors.

If an electrical current were created in a superconductor, it would last for hundreds of thousands of years. A superconductor, then, is a perfect conductor, and so the magnetic flux through a superconductor ring cannot change, which leads to *superconductor screening*. In a perfect conductor the magnetic flux is always the same as it was when the material made the transition to $\rho = 0$:

$$\rho = 0 \Rightarrow \frac{d\vec{B}_i}{dt} = 0.$$

An important consequence is that a perfect conductor keeps its magnetization constant. Even if the external magnetic flux density is switched off, the finite magnetic flux creates persistent currents that keep the magnetization constant.

A basic difference between a perfect conductor and a superconductor is that the latter shows the *Meissner-Ochsenfeld effect*. This effect causes a material in a superconducting state to always have zero magnetic flux density. That is, the superconductor expels magnetic flux density lines from inside:

$$\rho = 0 \Rightarrow \vec{B}_i = 0.$$

As a consequence, inside a superconductor the magnetic permeability is zero and the susceptibility corresponds to that of a perfect diamagnetic material. In a superconductor, then, the induced currents are only on the surface and have a penetration length of roughly 10^{-5} cm, depending on the temperature ($\lambda = \lambda_0[1 - (T/T_c)^4]^{-1/2}$).

There is a critical maximum value of the external magnetic field at which the superconductor stops being so ordered. The expression for the critical field is $H_c \approx H_0[1 - T/T_c]^2$, which also implies a critical value for the current density. We know that $T_c(H) < T_c(0)$; therefore, the magnetic field increases disorder in the superconducting state and allows superconducting systems to be cooled by a magnetocaloric effect, called adiabatic magnetization, similar to the adiabatic demagnetization for paramagnetic materials. This process entails:

1. Thermal insulation of the superconducting sample.

2. Adiabatic application of a magnetic field to the sample. In a strict sense, this is called adiabatic magnetization. As the magnetic field creates disorder in the superconductor, the entropy and magnetic energy increase; because of the insulation, the total energy remains constant, so the thermal energy and thus the sample temperature are lowered. This step can be explained from a magnetic point of view—that is, regarding superconductors as diamagnets.

3. Isothermal elimination of the magnetic field. This allows the sample to return to the initial state but at a lower temperature.

An energy band is associated with the superconducting state, corresponding to a two-level theory. Low critical temperature superconductivity can be understood as the result of electronic interaction via phonons. If $T < T_c$, these phonons will cause two electrons to interchange energy quanta in such a way that they join in a slightly bound state. This bound state, called a Cooper pair, is a boson (i.e., has an integer spin) and does not follow the Pauli exclusion principle. All the Cooper pairs can be in the particle ground state, which permits the resistivity to disappear. In this way one can sketch the energy band gap (figure 7.1).

A forbidden zone allows photons with energy greater than twice the forbidden zone width to break the Cooper pairs. This energy usually corresponds to infrared radiation. At $T > 0$ K thermal excitation complicates this interpretation.

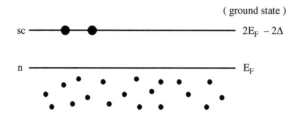

Figure 7.1 Scheme of the energy band gap for the ground state of a Cooper pair. "sc" indicates superconductor and "n" normal metal.

The isotope effect (also present in high critical temperature superconductors) is expressed as $T_c = f([isot])$; specifically $M^\alpha T_c = ct$, with $\alpha \approx 0.5$. This reinforces the argument that the electron-phonon interaction is important in superconductivity.

Junctions with superconducting materials display a characteristic tunneling effect, which we treat later on.

The main applications are clear:

- Exploitation of the use of a perfect conductor.

- Exploitation of the Meissner-Ochsenfeld effect (magnetic levitation).

- Exploitation of the adiabatic magnetization.

- Exploitation of the Josephson effect (see section 7.3).

7.2 TYPE I AND TYPE II SUPERCONDUCTORS

A possible classification of superconducting materials could be as high or low critical temperature superconductors. But this is a vague definition since "low" or "high" is arbitrary. A more precise classification is below.

Intermediate State

The magnetic flux density $\vec{B}$ inside a superconductor is zero; however, the magnetic field $\vec{H}$ may be different from zero due to surface electric currents that lead to a nonzero magnetization intensity. This implies the presence of a demagnetizing factor.

Consider a solenoid with a superconductor in it (figure 7.2). The applied magnetic field (created by the solenoid) will be H_a and N is the number of solenoid turns. Then

$$Ni = \oint \vec{H} \cdot d\vec{l} = \int_{AB} \vec{H}_i \cdot d\vec{l} + \int_{BCDEFA} \vec{H}_e \cdot d\vec{l}.$$

If we remove the superconducting sphere,

$$Ni = \oint \vec{H} \cdot d\vec{l} = \int_{AB} \vec{H}_a \cdot d\vec{l} + \int_{BCDEFA} \vec{H}'_e \cdot d\vec{l}.$$

Surface currents screen the field, so $H_e < H'_e$; consequently, $H_i > H_a$. As we see, the internal field is greater than the applied one in such a way that the critical state H_c is reached inside the superconductor for applied fields H_a weaker than the critical one. To summarize,

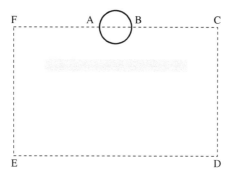

Figure 7.2 Illustration of field screening by surface currents on a superconductor (see text).

$\vec{H}_i = \vec{H}_a - n\vec{M}$, where n is the demagnetizing factor. As the superconductor shows perfect diamagnetism, $\vec{M} = -\vec{H}_i$ and

$$\vec{H}_i = \frac{\vec{H}_a}{1 - n}.$$

The superconducting state destabilizes when the internal field is greater than H_c, meaning that for an applied field value of $H'_c \equiv H_c(1 - n) < H_c$ the material can be neither in the pure superconductor state nor in the normal state, although in a normal metal $H_i \approx H_a = H'_c < H_c$ and thus the critical field is not reached.

Summarizing, at $T < T_c$ for a type I superconductor (depending on the sample shape),

- $0 < H < H'_c = (1 - n)H_c$ indicates a type I superconductor in the pure state,

- $H'_c < H < H_c$ indicates a type I superconductor in the intermediate state, and

- $H > H_c$ indicates a metal in the normal or nonsuperconducting state.

The superconductor in the intermediate state will have superconducting and normal zones (*coexistence region*). In this state the superconducting and normal zones are distributed in such a way that the field inside the superconducting zones is smaller than the critical one due to the surface currents; the field inside the normal zones is greater than the critical one. Thus, the superconductor fraction is x_{sc} and the normal one is x_n. The effective magnetic flux density is

$$\vec{B}_{eff} = \frac{x_{sc} \cdot 0 + x_n \cdot \vec{B}_n}{x_{sc} + x_n} \equiv \eta \vec{B}_n = \eta \mu_0 \vec{H}_i$$

where $\vec{M}_{eff} = \vec{B}_{eff}/\mu_0 - \vec{H}_i$ and therefore

$$\vec{H}_i = \frac{\vec{H}_a}{1 + n(\eta - 1)}.$$

In summary, the intermediate state appears in a type I superconductor for nonzero applied fields H_a in the interval $[(1 - n)H_c, H_c]$ (figure 7.3).

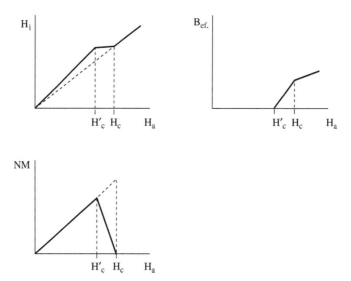

Figure 7.3 Magnetic fields and magnetization versus applied field in a type I superconductor.

Type II Superconductors

So far, we have discussed superconductivity based on electron-electron attractive interactions by way of phonons that produce bosons; these move easily in a perfect monatomic metal. The fact of having to cross an energy barrier in order to break the Cooper pairs leads to a *correlation length* ξ (also called the *Pippard coherence* length); ξ is the length at which the number of Cooper pairs fluctuates. In known cases the correlation length is large (around 1 μm). On the other hand, there is a *penetration length* of the magnetic flux density. Thus, surface currents screen the magnetic flux density inside the superconductor. But these currents are not strictly on the surface, penetrating the superconductor for a distance also called the *London length* λ. Clearly, $\xi \gg \lambda$. Yet if the superconductor has imperfections or is not a pure metal (i.e., it is an alloy), the coherence length is much smaller than in a pure and perfect metal. If $\xi < \lambda$, then the appearance of zones far from the surface (more than the order of magnitude of ξ), with a magnetic flux density $\vec{B}$ quite different from zero, is energetically favorable; that is, normal zones appear inside the superconductor. This type of superconductor is called type II (actually those with $\xi/\lambda < \sqrt{2}$). It is intrinsically different from the intermediate state of type I superconductors, which depends on the sample's shape. If the creation of normal zones inside the superconductor is favorable, then a *mixed state* appears. This mixed state consists of cylindrical normal zones called *fluxoids* where quantized electrical currents make the magnetic flux density greater than the critical value (figure 7.4).

Fluxoids are arranged in a geometrical way because of the repulsion between two parallel solenoids. Then, a lower and an upper critical field arise:

- If $H < H_{c_1} \approx H_c \xi/\lambda$ the material is a type II superconductor in the pure state.

- If $H_{c_1} < H < H_{c_2} = \sqrt{2}\lambda/\xi\, H_c$ the material is a type II superconductor in the mixed state.

- If $H > H_{c_2}$ the material is in its normal state or is a nonsuperconductor.

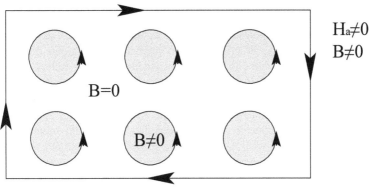

Figure 7.4 Sketch of surface currents and fluxoids in a mixed state.

For H_{c_2} a paramagnetic limit complicates the parallel disposition of the Cooper pairs' spins; thus, if H_{c_2} is very high it is reduced due to this paramagnetic limit. Note also that $H_c \in (H_{c_1}, H_{c_2})$ (H_c is called the *thermodynamic critical field*). The reason for this is that in the mixed state the free energy is smaller than in the pure state with the same parameters. Hence it is necessary to increase H_a more than H_c to have an unfavorable mixed state (figure 7.5).

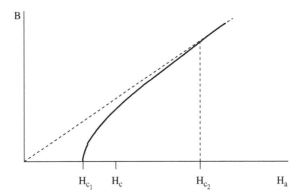

Figure 7.5 Magnetic flux density versus applied field in the mixed state.

This means that for $H_a = 0$ the T_c in type II superconductors is greater than in type I superconductors, as we find, for example, in the critical temperature of Nb$_3$Ge, which is 22.5 K.

High Critical Temperature Superconductors

High critical temperature superconductors do not fit well in either of the two types of superconductor described above. Even though there is not yet a theory for them, they are supposedly favored in some crystallographic planes. The most common (that is, cheapest) high critical temperature superconductors are ceramic materials of the perovskite type YBaCuO. The disposition of copper oxide planes, where superconductivity is favored, indicates that this kind of superconductivity has a key structural basis—even in the normal state the material has strong anisotropy in properties like magnetic flux density penetration length and conductivity. The values are two orders of magnitude higher in the copper oxide plane than in the perpendicular direction. The Cooper pairs still originate from strong coupling between electrons via phonons. In these instances the interaction is more complex and can be detected only by electronic velocity changes generated by the oxygen atoms' movement.

The critical currents of the superconductive state also manifest this anisotropy. A high critical temperature superconductor can be modeled as a quasi-2D superconductor that contains planes connected by weak junctions (see section 7.3); these junctions reduce the critical current in the direction perpendicular to the copper oxide planes.

Other perovskite-like superconductors are compounds of the type LaBaCuO and TlBa-CaCuO. In recent years we have learned that other isostructural compounds with other compositions also lead to high critical temperature superconductivity. For those of type YBaCuO, when in $YBa_2Cu_3O_x$ we have $x = 6$, we obtain insulating antiferromagnetic behavior; if we set $x > 6.5$, we get nonmagnetic metallic behavior added to a change in crystal structure; and if we set a value for x greater than 6.64, the material acquires superconductive behavior, until $x \approx 7$.

This kind of superconductor is more useful because the temperatures needed are higher than the boiling point of nitrogen. The main drawbacks are that the critical temperature is still far from room temperature and that their ductility is too low to make common electrical cables.

The mercury cuprates like $HgBa_2Ca_2Cu_3O_{8+x}$ have critical temperatures up to 134 K. The highest critical temperature expected in this type of cuprate is around 160 K, for the highly radioactive compound $HgRa_2Ca_2Cu_3O_8$.

There is also research on sodium-doped wolframates [18], where superconductive-like phenomena have a critical temperature higher than that of boiling liquid nitrogen. Here it seems that Cooper pairs are formed by phononic interactions, unlike the cuprates where the Cooper pairs are highly favored by magnetic interactions.

The common features of high critical temperature superconductors are that they are

- Dielectric materials that can be doped until they acquire electrical conductor properties,

- Materials in states near structural and electronic instabilities, and

- Materials having surface states.

7.3 JOSEPHSON EFFECT

Tunneling is the quantum effect that allows a particle to pass through a potential barrier that is higher than the particle energy, which is not classically possible. Assume that we have two superconductors separated by a vacuum; the stationary wave representing an electron in the vacuum decays exponentially as $e^{-x/\xi}$ where ξ is approximately 10^{-8} cm. Then, if the *forbidden* distance to be traveled by the electron is smaller than 10^{-7} cm, there will be a noticeable tunneling electronic current. Assume, then, that two conductors in a vacuum at $T = 0$ K are separated by a short distance. If both conductors are at the same electrical potential, the tunneling effect is not possible due to the Pauli principle, because the most energetic electrons of one of the metallic materials have, at most, the same energy as those of the other metallic material (the Fermi energy). But if we impose a potential difference between the two materials, a tunnel current will develop between them, because the conduction electrons will occupy free levels and will not contravene the Pauli principle.

If we assume that a metallic material and a superconductor in a vacuum at $T = 0$ K are separated by a tiny distance, we find a slightly different behavior. Here, for zero applied voltage, the energy of the Fermi level in the metallic material is the same as the energy of the Cooper pairs' condensed levels (condensed because there are many and the Pauli principle does not affect them). Hence there is no tunneling effect. If we apply a potential difference to the superconductor, there is no tunneling effect until V reaches the value Δ/e, where the lower part of the continuum of energy levels corresponding to the quasiparticles matches the Fermi level in such a way that the electrons can go from the metallic material to the superconductor by tunneling (figure 7.6).

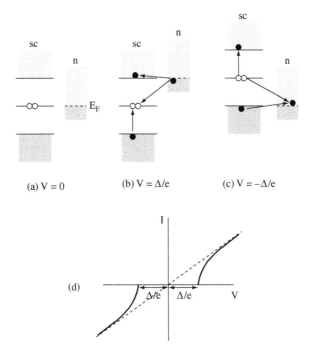

Figure 7.6 Tunneling effect at 0 K between a normal metallic material (n) and a superconductor (sc). (a), (b), and (c) are for different applied V and (d) is the I-V characteristic.

When the voltage is negative tunneling cannot occur until the voltage is $-\Delta/e$ and an electron moves from the occupied part of the electronic conduction band of the superconductor to the upper part of the conduction band of the normal metal. Also, a Cooper pair possibly breaks down and one of the electrons goes to the quasiparticle level (for this reason we need the potential energy) and the other one falls to the Fermi level of the metallic material. Note that when only one electron is involved, it is necessary for it to move horizontally on the energy scale to conserve energy. If there are two, one might increase and the other decrease its energy.

Assume that we have two identical superconductors at $T \neq 0$ K (that is, the quasiparticle levels are filled). If we do not apply any voltage there can be no tunnel current. If we do apply a voltage, from the first moment the excited particles could manifest a tunneling effect; but just a few particles can do this, and it leads to a nearly constant (when varying the voltage) and tiny current. This occurs as far as $V = 2\Delta/e$, where electrons from the filled conduction band of one superconductor can go to the quasiparticle levels of the conduction band of the other superconductor. Also, the disintegration of a Cooper pair (figure 7.7) is possible; each electron goes to one of the two quasiparticle levels.

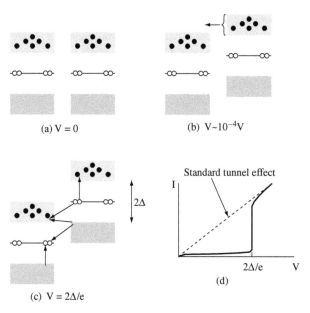

(a) $V = 0$ (b) $V \sim 10^{-4}V$

(c) $V = 2\Delta/e$

Figure 7.7 Tunneling effect between two identical superconductors. (a), (b), and (c) are sketches for various applied V and (d) is the I-V characteristic.

If the two superconductors are different, the I-V characteristic is also different (figure 7.8) but can be explained in a similar way. Note (figure 7.8) that there is a zone with a negative differential resistance.

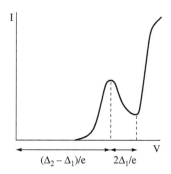

Figure 7.8 I-V characteristic of the tunneling effect between two different superconductors.

In addition to these tunneling effects, the *Josephson effect* is a tunneling effect of the Cooper pairs themselves. This effect is possible because condensation allows many Cooper pairs. It appears when there is an insulating thin film between two superconductors and there is no potential difference between them. This flux is called the *tunnel supercurrent*.

If the Josephson current density exceeds j_c a voltage drop V appears through the junction, indicating that the tunneling effect is simultaneous for electrons and Cooper pairs. In this situation, the Cooper pairs cannot conserve the energy by themselves. It is necessary to balance the energy by other means; in this case, a photon of frequency ν and energy $2Ve$ is emitted. This effect is called the alternating current Josephson effect. As $V \sim 10^{-3}$ V the radiation is in the microwave region. With measurements of the Josephson effect, superconducting energy band gaps can be precisely determined, as well as their changes with temperature or applied magnetic field or both.

The Josephson effect can be completely characterized by the difference of phase of the Cooper pair wave functions between one superconductor and the other. The Josephson supercurrent, that is, the electrical current of Copper pairs is $i_s = i_c sen(\Delta\phi)$, where i_c is the gap's critical current, and the voltage between the superconductors turns out to be $2Ve = \hbar(d/dt)\Delta\phi$.

Thus, we can draw an equivalent electrical circuit for a Josephson junction (figure 7.9). $I = C\,dV/dt + V/R + i_s$, where V/R is the electronic tunnel current, and $C\,dV/dt$ is the current caused by having a capacitor between the two superconductors. If we substitute the values as a function of the phase difference, we obtain an equation where the phase difference in a Josephson junction is governed in the same way as the angle in a forced and damped rigid pendulum:

$$I = \frac{C\hbar}{2e}\frac{d^2}{dt^2}\Delta\phi + \frac{\hbar}{2eR}\frac{d}{dt}\Delta\phi + i_c sen(\Delta\phi).$$

Thus, the frequency in an alternating current Josephson effect (in comparison with that of the pendulum) is

$$\nu = \frac{1}{2\pi}\left\langle\frac{d}{dt}\Delta\phi\right\rangle = \frac{2e}{h}\langle V\rangle = \frac{2e}{h}V_{dc}.$$

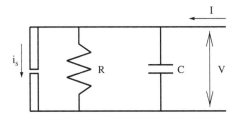

Figure 7.9 Equivalent circuit of a Josephson junction.

By way of summary, we now know that when a stationary current greater than the critical one passes through the Josephson junction a voltage V_{dc} appears. Yet in this instance the Cooper pairs tunneling effect leads to an oscillating component, whose frequency is $(2e/h)V_{dc}$, added to V_{dc}. Thus the Josephson junction can serve as a tunable oscillator given a voltage imposed at the junction. What is more, it is also a radiation detector, since electromagnetic radiation may lead to a V_{dc} in the system.

Weak junctions are generalizations of Josephson junctions. They can be point contacts or a narrowing of the superconductor. Their qualitative behavior is the same as that of Josephson junctions because in weak junctions the critical current is much smaller than in a bulk superconductor.

An important application of the Josephson effect is the SQUID (superconducting quantum interference device), which is a superconducting ring with one or more weak junctions.

Because the currents passing through the weak junctions are small, the linear momentum of the Cooper pairs will also be small; consequently, they will have a large wavelength, and if there is no applied field, the phase difference will be practically zero.

The fundamentals of a SQUID are as follows. If a superconducting ring is cooled without a field, and if a magnetic field is applied in the superconducting state, then a supercurrent i cancels the internal field in such a way that $L|i| = \Phi_a$, where L is the ring's self-inductance. If there are some weak junctions and the critical current is small, except for extremely small applied fields the current cannot be high enough to cancel the internal magnetic field. Also, the current cannot change the phase difference significantly. Most SQUIDs are built in such a way that $i_c L \ll \Phi_0$ (Φ_0 is the magnetic flux of a fluxon: the quantized magnetic flux). Then, $\Phi \approx \Phi_a$ and there cannot be an integer number of fluxons due to the weak junction.

We can show that the phase difference created by magnetic flux density is $\Delta\phi(B) = 2\pi \, \Phi_a/\Phi_0$. But the phase should change by an integral multiple of 2π in a superconducting ring, and so a current i will appear and cause the total phase difference (if there are m weak junctions) to be an integer multiple of 2π: $\Delta\phi(B) + m\Delta\phi(i) = 2\pi n$. In general, current variations should be measured when the magnetic flux is slightly varied such that the ring phase difference is a multiple of 2π. In that way, the current variations will have a period equivalent to the flux variation of one fluxon. This permits the use of the SQUID as a precise magnetometer.

PROBLEMS

7.1. What does it mean that a magnetic field disorders a superconductor? Why does it?

7.2. Explain a Cooper pair. Why is it an essential ingredient of superconductivity?

7.3. Why does the intermediate state appear in type I superconductors?

7.4. Interpret the I-V curve for the electronic tunneling effect (figure 7.8) corresponding to two different superconductors [16].

7.5. Explain the main differences between the intermediate state (type I superconductors) and the mixed state (type II superconductors).

7.6. Give examples of the differences and similarities between the intermediate state and the mixed state, from subjects of materials science other than superconductors.

7.7. What role do the coherence length and the penetration length play in shape aniso-tropies?

7.8. In the preceding problem, argue whether the mixed state can be considered a result of a shape anisotropy. Why?

7.9. Knowing that inside the fluxoid the sample is not a superconductor, argue why the fluxoids get pinned in crystal imperfections.

7.10. Calculate the dc voltage that appears at a Josephson junction of mercury at 0.2 K if it is "illuminated" with electromagnetic radiation of 483.6 MHz ($h = 4.136 \times 10^{-15}$ eV·s).

7.11. How does a SQUID with only one weak link work?

Chapter Eight

Optical Materials

8.1 THE INTERACTION OF RADIATION WITH MATTER

The interaction of radiation with matter is so prevalent that it could be regarded as a science unto itself. According to this view, all of optics, electromagnetism, and quantum mechanics come into play. Restricting ourselves to the optical region of the *electromagnetic spectrum* (figure 8.1), we take into account a zone where the radiation's wavelength is much greater (of the order of 10^3 times) than the crystal spacing and the size of molecules that compose solids. Hence here we do not consider phenomena like the diffraction of radiation by crystal structures described in section 2.6, because they require an optical wavelength of the same order of magnitude as the lattice constant of the crystal structure, as occurs in the X-ray region of the spectrum.

The energy of photons in the visible region ranges approximately from 1.8 to 3.1 eV. The microscopic processes involved in producing and detecting radiation in this interval of energy relate to the valence electrons of atoms. Molecular vibrations and rotations are responsible for phenomena in the near infrared range (lower energies) and the inner electrons in atoms or ions are responsible for phenomena in the ultraviolet range (higher energies).

After new detection devices and radiation sources like lasers appeared, the word "optics" came to be applied to the entire range including the near infrared and ultraviolet. In fact, detection and generation of radiation is performed now by similar devices in the whole region (figure 8.1), independently of whether the radiation can be detected by the human eye.

Semiconductors are used for both generation and detection at different wavelengths, and they are the basic materials of optoelectronics, the science where optical and electronic signals interact in communications systems, computers, or data processing. Optoelectronics has now moved its frontiers to the lower infrared and to the blue regions.

A common technology also simplifies calculations. Often, phenomena and properties can be treated by the same approximate mathematical approaches, usually derived from classical optics. For some special properties a *complete quantum theory* description is necessary. This means that, as in quantum optics, the matter properties and electromagnetic field are both quantized. Normally the simplifications, which in order of reducing complexity, are

- Semiquantum theory (quantum description of electromagnetic field, but classical description of matter);

- Semiclassical theory (classical description of electromagnetic field, but quantum description of matter); and

- Classical theory (classical description of both matter and electromagnetic field).

When a material interacts with radiation two main effects are produced: First, the matter is affected by the electromagnetic field changing its charge distribution; second, these effects on the material's charges or on the crystal lattice feed back to the radiation field.

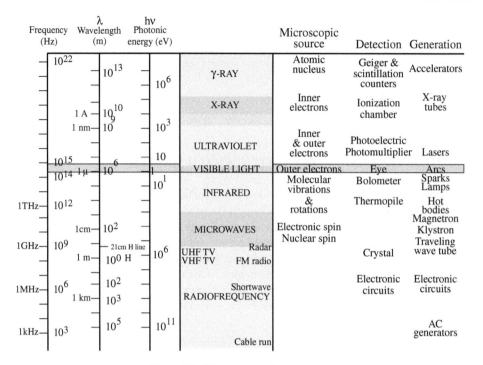

Figure 8.1 Electromagnetic spectrum.

Given the possibility that secondary effects may be neglected, we should use one or another simplification of the theory. In this chapter we do not consider the second effect.

The diverse physicochemical processes that radiation creates in crystalline solids are usually grouped according to the radiation's corpuscular character and whether the photon energy $E = h\nu$ is conserved. In this context, energy conservation implies that the radiation's *color* remains fixed, a phenomenon that happens if energy is not transferred to the solid.

When the photon energy is conserved (table 8.1) the phenomena are those of classical optics. As a first approximation, matter can be described by macroscopic constants like the *refractive index*, without bringing in quantum effects.

Table 8.1 Radiation-matter interaction (interactions where radiation coherence is not important nor does the field intensity produce nonlinear effects)

Energy of Photon Conserved	Energy of Photon Not Conserved	
Phenomena of Wave Optics	Electrical Phenomena	Nonelectrical Phenomena
Reflection	Photoelectric emission	Photoluminescence
Transmission	Generation of a free electron	Heat production (phonons)
Rayleigh scattering	Generation of an electron-hole pair	Exciton creation, Raman scattering

Detailed studies of radiation absorption or emission phenomena including charge interaction, scattering, and change in frequency require that matter be considered in a microscopic and quantum way. In these phenomena, the photon energy is not conserved (table 8.1). These phenomena are normally grouped into two subfamilies: those whose effects are due to electrical charges, and those where nonelectrical phenomena generate photons with different energies, phonons, or excitons.

In this chapter we dwell on phenomena whose energy does not conserve the incident photon's energy, and we also reexamine some topics of optics and electromagnetism. At the end of the chapter we will consider the phenomenon of polarization because sometimes it enables us to link optical properties with mechanical ones, such as internal states of stress, or with electrical ones, like anisotropies induced by stresses or electrical fields.

8.2 OPTICAL PARAMETERS

When macroscopically characterizing the propagation of an electromagnetic wave in an optical material via classical methods, the refractive index is widely used. The macroscopic phenomena that can affect an electromagnetic wave that comes from a vacuum and penetrates a material medium are *reflection*, transmission or *refraction*, *absorption*, *scattering*, and *polarization modification*.

Electromagnetic waves propagate in a vacuum—far from intense gravitational fields—at a speed of $c = 2.99792458 \times 10^8$ m $\cdot$ s^{-1}. This quantity, a convention, allows one to determine the meter's measurement standard. Classical electromagnetism establishes $c = (\epsilon_0 \mu_0)^{-1/2}$, where $\mu_0 = 4\pi \times 10^{-7}$ N $\cdot$ A^{-2} and $\epsilon_0 \approx 8.85 \times 10^{-12}$ (A $\cdot$ s)$^2 \cdot$ N$^{-1} \cdot$ m^{-2}. The value of c can be experimentally obtained by making electromagnetic, optical, or astronomical measurements. It unifies electromagnetism with optics, and thereby has a significant role in the history of science.

In a material light propagates at a speed v slower than the speed of light in vacuum. This speed defines the refractive index n: $v = c/n$, where $n = \sqrt{\epsilon_r \mu_r}$. The preceding formula is called the *Maxwell relation* when it is taken for granted that the material is not magnetic (and therefore $\mu_r = 1$, valid for transparent optical materials). The refractive index depends, of course, on the radiation frequency (table 8.2).

The response of materials to the variable electrical and magnetic fields of electromagnetic waves changes according to what type of material we consider. So, in a metallic material, electrons in the conduction band are basically free and respond immediately to the fields, thereby producing electrical currents formed by accelerated electrons. The Maxwell equations (see chapter 6) establish that the accelerated electrical charges emit electromagnetic radiation, resulting in a perfect conductor that reemits radiation. In an imperfect conductor made of metallic materials, the accelerated conduction electrons lose part of their energy in an irreversible way, transforming it into heat by the Joule effect. This effect of finite conductivity (see section 4.1) generates a major attenuation. In dielectric materials, the valence electrons bind to ions in such a way that the variable fields create the appearance of oscillating electrical dipoles; their electrons have resonance frequencies with maximum amplitude.

The fact that there are no perfect insulators means that the permitivity must have an imaginary component (see section 6.5). This makes the refractive index complex, $\tilde{n} =$

Table 8.2 Maxwell relation (at 60 Hz) and refractive index (at 5×10^{14} Hz) for some materials at room pressure

Material	$\sqrt{\epsilon_r}$	n
Gases at 0°C		
Air	1.000294	1.000293
He	1.000034	1.000036
H_2	1.000131	1.000132
CO_2	1.00049	1.00045
Liquids at 20°C		
Benzene	1.51	1.501
H_2O	8.96	1.333
Ethanol	5.08	1.361
Carbon tetrachloride	4.63	1.461
Solids at 20°C		
Diamond	4.06	2.419
Amber	1.6	1.55
Fused silica	1.94	1.458
NaCl	2.37	1.50

$n - i\kappa$, with the real and imaginary parts linked by the Kramer-Krönig relation. The real part is the classical refractive index; the imaginary part is the *extinction coefficient*, which is related to the material's conductivity. Thus, if the propagation of a monochromatic plane wave (generally a solution of Maxwell's equations) through a material has a refractive index of $\tilde{n}$, the electrical field can be expressed as

$$\vec{E} = \vec{E}_0 e^{\iota\omega\left(t - \tilde{n}\ \vec{r}\cdot\hat{v}/c\right)},$$

where $\hat{v}$ is the unit vector in the direction of the radiation propagation. If we introduce the expression for the complex refractive index, we obtain

$$\vec{E} = \vec{E}_0 e^{-\kappa\omega\ \vec{r}\cdot\hat{v}/c} e^{\iota\omega\left(t - n\ \vec{r}\cdot\hat{v}/c\right)},$$

where the first exponential represents a damping of the wave and the second is the wave's progression.

The magnitude normally detected is the intensity of the wave, proportional to its energy and therefore to the squared modulus of the electrical field:

$$I = I_0 e^{-\alpha\vec{r}\cdot\hat{v}},$$

where $\alpha = 2\ \omega\kappa/c = 4\pi\ \kappa/\lambda_0$ is called the *absorption* or *attenuation coefficient*.

The curve showing the value of α as a function of energy ($h\nu = \hbar\omega$), or as a function of the radiation wavelength in vacuum, λ_0, is the *absorption spectrum* (figures 8.2 and 8.3).

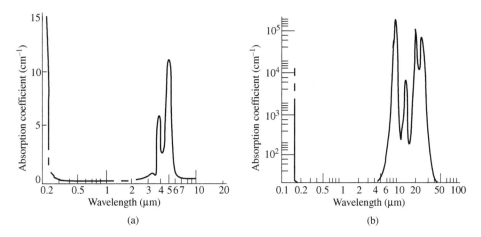

Figure 8.2 Absorption spectra in dielectrics; (a) pure diamond and (b) quartz.

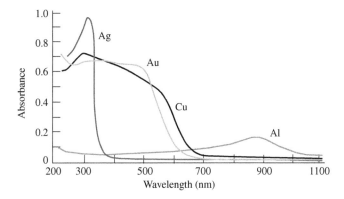

Figure 8.3 Absorption spectra of silver, gold, copper, and aluminum.

8.3 PROPAGATION OF LIGHT IN SOLIDS

Propagation of Electromagnetic Waves in Dielectrics

To see how the real part of the refractive index varies according to frequency $n(\omega)$ or to its wave vector $n(\vec{k})$ (directly linked to the *dispersion relation* $\omega(\vec{k})$), the absorption of energy by the dipoles must be taken into account as demonstrated by the polarizability (section 6.5). Hence the real part of the refractive index behaves similarly to the real part of the permittivity (figure 6.10).

In regions of low absorption (transparent regions) the refractive index increases slowly when the frequency rises (or decreases when the wavelength λ_0 rises) until it reaches an absorption band (a resonance line) where the refractive index diminishes—a zone that manifests *anomalous dispersion*.

Propagation of Electromagnetic Waves in Conductors

In a metallic material the conductivity is different from zero; this implies that the absorption coefficient plays a valuable role in wave attenuation. When entering a metallic medium the initial intensity of a wave declines by a factor of e (the base of the natural logarithm) when the wave in the material has moved a distance $\delta = 1/\alpha$, called the penetration or skin depth. The material is transparent to monochromatic radiation if the skin depth is much longer than the sample's macroscopic length. Metallic materials have a non-null conductivity and thus the skin depth is small.

For example, copper in the ultraviolet zone ($\lambda_0 \sim 100$ nm) has a skin depth of $\delta = 0.6$ nm. The skin depth in the infrared ($\lambda_0 \sim 10,000$ nm), although longer, is still small, $\delta = 6$ nm. This is consistent with the fact that metallic materials are opaque, although they can become transparent in thin layers, such as the coatings used in partially silvered mirrors. The look familiar in metallic materials is brightness, due to the high reflectance that happens when an incident wave cannot penetrate them. A few electrons *detect* the transmitted wave, and, although each of them strongly absorbs it, the absorbed energy is dissipated. Finally, most of the energy appears as a reflected wave. Hence, most metallic materials, even ones with little in common, have a gray, silvery color that reflects incident white light regardless of the wavelength; they do not exhibit coloration. Also, the transmitted wave in metallic conductors has an electric field component in a direction that does not exist in vacuum.

A model that represents a material as a continuous medium, as was done here, works well in the infrared region, where wavelengths are long and frequencies low. When the wavelength decreases, the particulate nature of matter becomes more evident, and the visible and ultraviolet regions are not suitably represented by this model, principally when absorption and emission of radiation is involved. Instead, we should use classical atomic models, like Drude's (chapter 4) and other similar ones. These models provide suitable results, even without being complete quantum theories; the metallic material is a lattice of ions, having inner bonded electrons, and a free electron gas. The bonded electrons have resonance frequencies as in dielectrics, and free electrons manifest reflection independent of the wavelength. With this kind of model a dispersion relation similar to the dielectric one can be obtained, in which the electron gas and the bonded electrons make independent contributions. One indication that establishes the value of this relation compared to other simpler ones consists of looking at whether the material has coloration. If it does, this indicates the importance of bonded electrons in a material that selectively reflects and absorbs (figure 8.3).

As practical examples, gold and copper have a yellowish-red coloration because their reflectivities are higher at those frequencies; for smaller frequencies in the visible range gold appears yellowish. A thin layer of gold (as thin as the skin depth, thereby allowing transmission) transmits mostly a green-bluish light. The material appears with these colors when it is observed in relation to its transmission and when the material is illuminated with white light.

Reflection and Transmission Coefficients

For a wide variety of applications we need transparent or translucent materials (i.e., materials for which the fraction of transmitted energy is much bigger than the fraction absorbed or reflected, in the spectrum's optical region).

As was said above, metallic materials, except for thin layers which are transparent, are opaque. Yet the transparency that we will investigate is independent of the physicochemical properties of absorption common to each atom, because all the resonance frequencies fall outside the spectrum's optical region.

In these conditions, when a wave impinges on a transparent solid and penetrates it, part of its incident light intensity (flow of light energy) I_0 goes into the solid and another part I_R is reflected by the interface of the two media. If of the energy that flows into the solid part I_T is transmitted and part I_A is absorbed, the conservation of the total energy requires that $I_0 = I_R + I_A + I_T$. If we divide throughout by the incident intensity we obtain $1 = R + A + T$, where R is the *reflectance* or the coefficient of integral reflection, A is the *absorbance* or the coefficient of integral absorption, related to the coefficient of absorption α (and hence put in relation to $e^{-l\beta}$, l being the sample's thickness), and T is the *transmittance* or the coefficient of integral transmission. Each indicates the fraction of energy that is reflected, absorbed, and transmitted in an integrated way; that is, they are valid for the entire light bandwidth although their proportions vary according to the incident frequency or wavelength (figure 8.4).

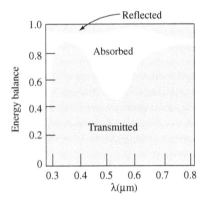

Figure 8.4 Light energy balance in a greenish glass as a function of the incident wavelength.

The proportions exhibited by these losses at different wavelengths induce transparent materials to reveal their coloration (the greenish color of window glass). If the absorption is uniform at all visible wavelengths, the material appears colorless, like the quality glasses used in optics, diamonds, or extremely pure sapphires. This behavior—really a simple consequence of energy conservation, giving no gain or loss of intensity at certain frequencies—is called the *Beer-Lambert law*.

Reflection, whose coefficient also derives from electromagnetism, is sensitive to the angle of incidence and the light's polarization. Thus, the *Fresnel coefficients* yield reflection values for the fields' amplitudes as a function of the incidence angle. In figure 8.5 we can see that the coefficients' values in extreme cases are the same as the ones when the electrical field's ($\vec{E}_0$) polarization components are parallel ($\parallel$) and perpendicular ($\perp$) to the radiation's surface of incidence. Since the biggest difference between a dielectric and a conductor happens in the latter instance, the reflectance at that angle is minimal; the Brewster angle does not reach a zero value as it does in a dielectric.

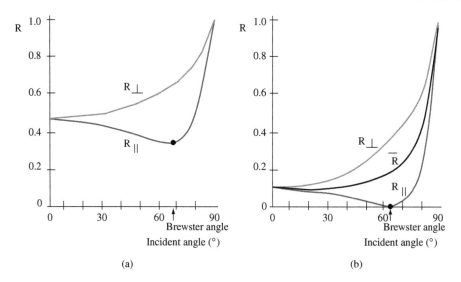

Figure 8.5 Reflection coefficients for (a) a metallic and (b) a dielectric material.

8.4 ABSORPTION AND EMISSION OF RADIATION IN SOLIDS

The processes that happen in absorption must be considered in detail, either when the entire absorption spectrum of a crystalline solid is studied, or when phenomena with electrical charges or with photons of different energies are added.

The opposite phenomenon to absorption is *stimulated emission* (not spontaneous emission), a new emission mechanism postulated by Einstein at the beginning of the twentieth century to explain using elementary physicochemical processes the Planck law for black-body radiation.

If we evaluate the processes correctly, the absorption is intrinsically stimulated; that is, it requires an electrical field to which it is proportional. *Spontaneous emission*, much easier to observe, is simply the excess of energy provided by an excited atom when it returns to equilibrium (the ground state). Stimulated emission processes are now quite significant. They are the key to *coherent radiation* or *lasers*, to the technological art of new industries, including telecommunications, home devices like the compact disk and DVD, and so on.

Let us first analyze the mechanisms appearing during a crystal's wave absorption in the spectral regions:

1. Absorption by electronic promotion from the valence band to the conduction band (section 2.5).

2. Absorption by excitons (section 3.2).

3. Absorption by free charge carriers (intraband transitions).

4. Absorption by imperfections (chapter 3).

5. Absorption by lattice vibrations (phonons) (sections 2.5 and 5.2).

The absorption coefficient α is associated with the probability of absorbing a photon of wavelength λ_0 in a sample of uniform thickness. If the sample is a crystal, diverse mechanisms come into play independently of each other; the coefficient of absorption, then, is the total of the absorption coefficients for each of the mechanisms described above. Regardless of the spectral region or the mechanisms involved, we know that the energy and the linear momentum must be conserved. Using a quantum mechanical description of systems we can determine the conditions for the existence of photons (quanta of light energy) and phonons (section 5.2). Phonons must participate in the absorption of radiation by electronic promotion: the variation of the electron's linear momentum in an indirect transition often cannot be compensated by the photons' linear momenta. This is the reason why the crystalline lattice has to give part of the linear momentum necessary for conserving the total in the form of phonons.

Elementary Absorption, Spontaneous Emission, and Stimulated Emission

Isolated Atoms

For a photon to be absorbed by an isolated atom, the incident photon energy $h\nu$ must be approximately equal to the difference of energy between the atom's two levels, $\Delta E \approx h\nu$. This relation is approximate owing to the Heisenberg principle, collisional broadening, and the Doppler effect (figure 8.6).

Figure 8.6 Absorption of a photon by an idealized isolated atom (two levels).

If the energy absorbed is not enough to excite an atom's electron to the continuum of levels where an electron can separate from others (the energy of ionization), the atom remains for a time in this *excited state*. After that, it decays to its *ground state*, or to its lower energy, recovering the incident photon energy by spontaneous emission of one photon if the transition is direct, or by several photons if the transition happens at intermediate energy levels, which can sometimes lead to nonradiative events. If the atom decays to the same energy level, the emitted photon has the same energy as the absorbed one (figure 8.7).

Figure 8.7 Spontaneous emission of a photon by an idealized isolated atom (two levels).

If the atom in the excited state interacts with another photon of the frequency corresponding to the energy interval, there is a possibility, different from zero and proportional to the number of photons present, that this photon will stimulate the atom's deexcitation and produces the stimulated emission of another photon with the same frequency and in phase with the first one (figure 8.8). This mechanism enables one to amplify the light and to build coherent radiation sources such as lasers.

Figure 8.8 Stimulated emission of a photon by an idealized isolated atom (two levels).

Metallic Materials

In a metallic solid, the conduction band is partially full (figure 2.16). The photons corresponding to visible radiation, between 1.8 and 3.1 eV, excite the electrons toward unoccupied energetic states above the Fermi level (figure 8.9). Apart from the case of an isolated atom, there are always empty levels that allow electronic transitions. This means that all optical photon energies coincide with an allowed energy value, which favors the photon's absorption.

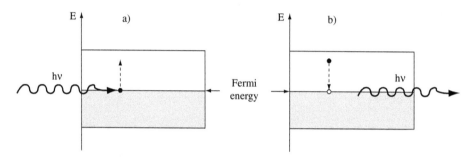

Figure 8.9 Absorption (a) and emission (b) by a metallic material.

As we mentioned in our electromagnetic approximation, in metallic materials this situation is not restricted to the optical region. All low-energy photons can be absorbed from radio frequencies up to the near ultraviolet. These materials are partially transparent to radiation at higher frequencies.

The energy is absorbed in a thin layer. It is reemitted at the same frequency as reflected light. The small portion of absorbed energy is transformed into heat produced in the decay of the excited electron.

Dielectrics and Semiconductors

DIRECT TRANSITIONS, $\Delta \vec{k} = 0$

In an insulator or intrinsic semiconductor the absorption of a photon provokes a promotion of electrons from the valence band to the conduction band. As a consequence, a hole appears in the valence band. If an electrical field is applied to the crystal, the free carriers (electrons and holes) will move in an electrical current, a phenomenon called *photoconductivity* (section 2.5). If the material has a band gap of magnitude E_g, the photon energy $h\nu$ that must be absorbed has to be higher than the band gap's energy, $h\nu \geq E_g$.

When electronic promotion is finished, the hole and its corresponding electron can either be free in the valence and conduction bands, respectively, or be in bonded states like excitons (section 3.2).

In the optical region of the spectrum where photons have energies between 1.8 and 3.1 eV, the gap of the forbidden zone must be smaller than the incident photon energy. If this does not occur, the electron goes back to the band it departed from and the solid does not absorb this energy. Thus, the maximum band gap for absorption of a visible photon is 3.1 eV. Above this minimum, all photons of higher energy are absorbed in an intense and continuous absorption band.

The edge of the absorption band tends to be approximate, with respect to the difference of energies. Most semiconductors have an absorption band in the spectrum's infrared region. Materials that have forbidden zones larger than this energy will be transparent (if they do not contain imperfections), and semiconductors that have smaller forbidden zones than this absorb all the energy of the incident light by electronic promotion and are opaque. Materials appear colored, have imperfections (defects and/or impurities), or intermediate forbidden zones. The usual experimental procedure for finding the magnitude of the forbidden zone (band gap), such as with indium antimonide, is to determine the mean slope of the absorption curve and where it crosses the energy axis.

The opposite process is the return of an excited free electron from the conduction band to the valence band, causing the disappearance of a hole (figure 8.10), called *recombination* and the appearance of a photon corresponding to the energy interval. The electron goes to the lower part of the conduction band and the hole moves to the higher edge of the valence band in a short relaxation time, between 10^{-12} and 10^{-10} s. The luminescence coming from the recombination has an emission curve whose width is limited to the low-frequency zone by the relation $h\nu = E_g$.

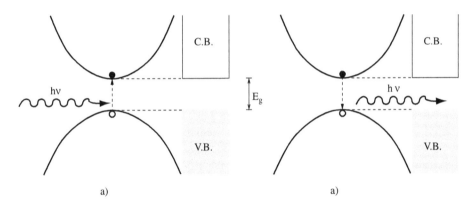

Figure 8.10 Direct absorption (a) and emission (b) for a dielectric or an intrinsic semiconductor. C.B. and V.B. indicate the conduction and valence bands, respectively.

INDIRECT TRANSITIONS, $\Delta \vec{k} \neq 0$

In semiconductors with more complex energy bands (chapter 2 and figures 2.13, 2.14, and 8.11), apart from direct transitions ($\Delta \vec{k} = 0$), indirect transitions are possible with variation of the electron's wave vector ($\Delta \vec{k} \neq 0$). In these instances additional mechanisms allow the conservation of the total energy and linear momentum, as with the appearance of a phonon of energy E_F. The energy that the photon needs to move an electron from the valence band to the conduction band is $h\nu \geq E_g + E_F$ if a phonon of energy E_F is emitted and $h\nu \geq E_g - E_F$ if a phonon of energy E_F is absorbed.

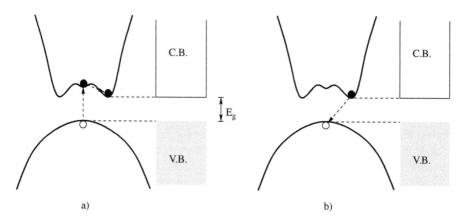

Figure 8.11 Indirect absorption (a) and emission (b) for a dielectric or an intrinsic semiconductor. C.B. and V.B. indicate the conduction and valence bands, respectively.

The probability of such a phenomenon is much smaller than that of a direct transition, although the energy difference is smaller because more particles are involved (electron, photon, and phonon, for example) to ensure conservation of energy and momentum. Materials such as Ge and Si have this type of band.

PRESENCE OF IMPERFECTIONS

When impurities (dopings) or defects (either at the surface or in the substrate) occur in semiconductors or dielectrics, localized states bonded to these imperfections appear in the forbidden zone. These localized states play a major role in absorption and emission, which is linked to the emission or the ionization of impurity centers (figure 8.12).

As seen in chapters 2 and 3, the presence of imperfections generates extrinsic levels, deep and shallow, that create absorption bands. For example, in an N-type material the electrons of the donor levels can be excited to the conduction band. These localized states also cause luminescence when electrons excited from the conduction band to the donor levels neutralize them and radiation appears in the far infrared.

If donors or acceptors are introduced in small proportions, the extrinsic levels in the forbidden zone create radiation only seen at low temperatures; when temperatures are high, the exciting mechanism of electronic thermal promotion prevents electrons from decaying to the extrinsic level. Because the energy of the extrinsic levels is within the forbidden zone, the absorption bands are located beyond the edge of the intrinsic absorption band.

Decay can also be produced by emitting two photons (figure 8.12), one during the electron's transition from the conduction band down to the extrinsic level, the other during the transition from the extrinsic level down to the valence band. As an alternative mechanism there is the possibility that photon emission dissipates the associated energy as heat.

Doping with impurities in small amounts can also determine the color of materials that are colorless in their pure state. For example, pure monocrystalline aluminum oxide Al_2O_3 is colorless. Once doped with between 0.5% and 2% of Cr_2O_3, the Cr^{3+} ion substitutes for the Al^{3+} ion in the crystalline structure of the material commonly called ruby. In the forbidden zone energetic levels that provoke strong absorption peaks in the blue-violet and green-yellow regions of the spectrum give rubies their reddish coloration (figure 8.13).

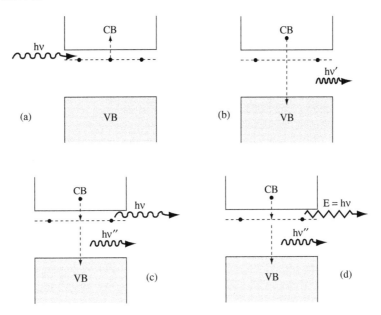

Figure 8.12 Transition through defect levels: (a) absorption, (b) photon emission, (c) two-photon emission, (d) emission of a photon and a phonon. C.B. and V.B. indicate the conduction and valence bands, respectively.

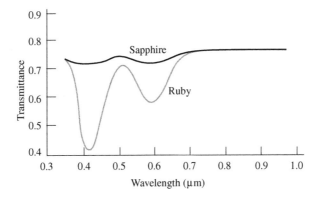

Figure 8.13 Transmission curves for ruby and sapphire.

The ruby has historical significance because it was used to build the first laser, although its value has decreased in comparison to newer solid-state lasers such as the neodymium family either in a glass or in an yttrium aluminum garnet (YAG) matrix, the titanium-sapphire laser, or alexandrite lasers. Gaseous lasers such as helium-neon or argon ionized lasers have also become more important.

Materials doped with other oxides that introduce ions as in inorganic glasses are used for the manufacture of colored glasses. They are employed in glassware and building and decoration.

LUMINESCENCE

We have defined *luminescence* as the reemission of visible light that follows energy absorption, and we have noted the diverse mechanisms that produce it. The energy normally comes from electromagnetic radiation of higher energy (e.g., the ultraviolet) that causes transitions from the valence band to the conduction band, or in other energetic forms: collisions with high-energy electrons, heat, mechanical stresses, or chemical reactions. The appearance of carriers in the presence of fields prompts photoconductivity that can be used to build detectors of light or photometers that measure the incident light intensity.

The events that occur until the energy is reemitted can make the luminescence fast or slow. When the luminescence lasts up to a second it is called *fluorescence*. When it lasts longer, it is called *phosphorescence*. In general, pure materials do not manifest these phenomena; they occur in materials that contain impurities in controlled amounts, like sulfides, oxides, wolframates, and some organic materials.

Luminescence has many applications. The most widespread is possibly the fluorescent tube. This tube contains mercury gas at low pressure where an electrical discharge generates radiation, mainly in the ultraviolet region. To be used as visible radiation, the glass tube's inner layer must be covered with silicates or wolframates that absorb the ultraviolet and transform it into visible white light.

Another important application is the cathode ray tube used widely in televisions, computer monitors, and oscilloscopes. Here an electron beam impinges on a screen which, because of a coating inside the device, converts electrical energy into visible radiation.

It is also possible to use the fluorescence properties of multilayer systems to obtain high capacities of information storage. In this way a compact disk (fluorescent multilayer disk, FMD) of CDROM size can store up to 140 Gbits. Here the information is not obtained via reflection, as with common CDROMs, but via incoherent light reemitted by different layers when they are illuminated by a laser beam.

In semiconductors, P-N junctions (section 4.3) are used as light emitters when strong electron-hole recombination is produced by a direct voltage through the device. Radiation can be obtained in the infrared or visible range, and nowadays in the near ultraviolet if suitable semiconductors are used. These diodes are known as light-emitting diodes (LEDs). Some digital screens and different displays for instruments use them. The color of emission of the LED depends on the semiconductors that were used to make it.

Finally, sunlight energizes *photovoltaic cells* to charge batteries in order to generate clean energy or to supply small instruments directly. Their operation is the inverse of the LED function. If a P-N junction is used, photoexcited electrons and holes leave the junction in opposite directions and create part of a current whose circuit is closed externally by a load. In construction of these devices, silicon is used more than other materials because it is so abundant in the earth's crust. As highly pure monocrystalline silicon is expensive, hydrogenated amorphous (chapter 9) or polycrystalline silicon is used. Its cost, its insensitivity to the presence of impurities, and its stability make it suitable for industrial purposes.

8.5 LASERS

Coherent Radiation

So far, in all types of emission the appearance of each photon is spontaneous (i.e., independent of other photons); therefore, photons appear in all phases distributed randomly, having all the polarization components, and lacking any correlation among them. If we consider the distribution of all the emitted frequencies, except those in which pure states are involved, giving line spectra, it is normally a wide band frequency distribution comparable to the absorption distribution. This type of natural radiation is known as incoherent radiation.

The interest in having linewidths as narrow as possible—with phase-correlated photons and with a unique polarization defined by the group of emitted photons—has led since 1960 to the development of coherent radiation emitters or *lasers*. The laser does not really amplify light; instead, it is an oscillator, a device that can take energy from a continuous or alternating source of radiation, create an oscillation of pure frequency, and maintain its phase and frequency over a long distance. This distance is named the *coherence length* and the radiation is called *coherent radiation*.

These properties, easily obtained in the radio frequency region, are hard to find in the visible region and very hard at shorter wavelenghts. In fact, from the time of Einstein's prediction to the laser's fabrication, half a century had elapsed. Many applications use the laser's properties: monochromaticity, directionality, coherence, and polarization (optional). Additionally, high peaks of power, short pulses (up to 10^{-15} s for high energies), and other similar properties can be obtained.

The elements needed to build a laser—although their practical implementation differs according to the type of laser—are the following:

- A *pumping source* to raise atoms or molecules to an excited energy level. It can be a gaseous discharge, a source of incoherent white light, electrical currents, sunlight radiation, a nuclear explosion, and so on.

- An *active medium* in which more atoms are produced in the excited state than in the ground state; a situation called *population inversion*. In population inversion there is a privileged direction of space where light can be *amplified*. This is called the gain direction.

- An open *electromagnetic cavity*, that has only one stationary wave condition in the high Q direction. It is normally built with two mirrors of high precision with controlled reflectivities. This cavity is placed with its direction of high Q matching the gain direction.

Solid-State Lasers

As noted when we described absorption and emission in an isolated atom, stimulated emission permits light amplification. If an atom interacts with a photon whose frequency is related to the energy difference between two levels allowed for an optical transition, the incident photon stimulates the atom's decay to its ground state. In this situation another photon is created with the same frequency, phase, polarization, and direction as the incident one. Hence, the radiation comprised of both photons is coherent. To obtain coherent radiation, it is sufficient to prepare many excited atoms in a particular direction in space. This is achieved usually by making an active medium with a cylindrical shape where the size in one direction is much larger than it is in the other two directions. A bar of excitable material having typical dimensions of 1 cm in diameter and 15–20 cm length is appropriate for building a ruby laser.

Although in the idealized diagram of the isolated atom there are only two levels, it is necessary to involve at least three energy levels because stimulated emission is as probable as absorption; consequently, with two levels it will never be possible to generate population inversion to achieve gain in the sample (figure 8.14).

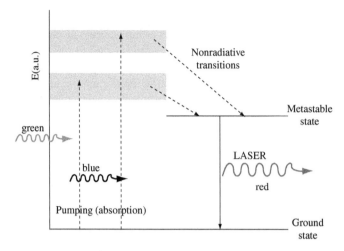

Figure 8.14 Energy levels in ruby.

Ruby's transmission curve (figure 8.13) enables us to observe that its greatest absorption is at frequencies in the blue-violet and green regions of the spectrum. Therefore, to excite these atoms by optical means it is convenient to illuminate them with a high-intensity source (the most suitable is the light emitted by a flash of a xenon lamp), where the emission is close to this region of absorption. Such illumination populates the levels (figure 8.14) and coherent radiation is obtained from Cr^{3+} ions that are excited. This is known as *optical pumping*. To obtain laser emission, enough gain must be achieved by means of an optical electromagnetic cavity. Two mirrors are placed at the ends of the ruby bar, so the initial photons can provoke a large amplification of coherent radiation that can remain inside the cavity for an indefinite time, traveling forward and backward.

If one of the mirrors is not a perfect reflector (for example, transmitting 10% and reflecting 90%), of each 100 photons that get to it, ten will go outside as useful coherent radiation. The 90 left will continue to amplify by stimulated emission again until in the next trip 100 photons reach the partially reflecting mirror. Because of the amplification a stationary condition is reached, giving a useful output of 10% of the energy stored in the cavity. The ruby laser has pulsed emission at 694.3 nm and the pulse's duration is slightly shorter than the duration of the flash lamp's shot. The lamp initially had a cylindrical shape and was placed at one of the focal points of an ellipse with the ruby bar at the other focal point. Later, helical or hollow shapes, or both, surrounded the ruby bar (figure 8.15) to illuminate it as efficiently as possible.

The structure of optical pumping in the ruby laser is also used in neodymium lasers. The ruby bar is replaced by a glass or a YAG crystal doped with neodymium ions (Nd^{4+}). The emission of neodymium lasers is in the near infrared (1.04/1.06 μm) and they are very useful lasers with many applications in material processing. Currently, neodymium-YAG lasers are the most used. Their frequency can be duplicated easily by generating second harmonics of the infrared radiation in a nonlinear crystal placed intracavity. Green emission

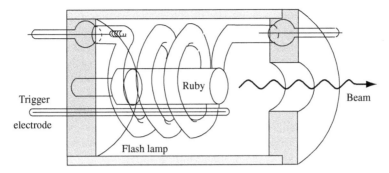

Figure 8.15 Sketch of a ruby laser.

(530 nm) is obtained, and it has become a standard of frequency. Neodymium-YAG lasers currently are pumped by LEDs and they are widely used in industry and medicine.

Other kinds of solid-state active media are different crystals used to obtain tunable lasers in the near infrared region like titanium-sapphire, alexandrite, and other similar ones. Normally they are pumped by a frequency-doubled neodymium-YAG laser.

Disordered media have coherent backscattering because there is symmetry in how the light scatters. Therefore, if the incident light is a plane wave, the diffused light in one direction interferes with light diffused along the other direction of the same path. This interference is constructive in the direction opposite to the illumination. Thus, if we add the contributions from each possible path, we derive a maximum of intensity in the direction opposite to the illumination. Meanwhile, some contributions are canceled out. This phenomenon can be observed in media like white sheets of paper, sugar, salt, and other disordered systems of white color that diffuse light well. This is the first step in building random lasers, where the active medium's gain and the pumping source combine their disorder. These lasers are generally built from monocrystals of pulverized titanium-sapphire. They are still the object of active research. The coherent backscattering phenomenon has the same foundation as do loops of light in disordered media, where light travels in closed paths in a disordered material. To have loops it is necessary that the anisotropy of the monocrystalline material be large enough for it to curve the light paths and for the disordered material to trap the light. These loops of light can give rise to photon localization in the same way that noncrystalline materials have localized electrons.

Semiconductor Lasers

We have seen that a P-N junction of some semiconductors can produce a high light intensity when a direct current is imposed that is capable of generating a high rate of recombination. Recombination produced in the depletion layer of semiconductors of the family called the *III-V group*, like GaAs, is balanced by emission of a photon. Other semiconductors like germanium or silicon mainly balance this energy by heat (vibrations of the crystal: a phonon) and clearly are not useful for luminescent diodes. From the IIIa group Al, Ga, and In are used and from the Va group N, P, As, and Sb.

Among them, there are two important families, one based on GaAs and the other on InP, including other elements of both groups to modify some properties like wavelength (e.g., GaN). A direct voltage produces carriers that are able to recombine in population inversion conditions, and if enough suitable in-phase photons appear in the region where the recombination is already occurring (figure 8.16), it is possible to obtain coherent radiation. This coherence appears if the gain is greater than the losses. When this condition is satisfied, the width of the emission frequencies narrows and the power at each frequency increases and reaches a peak in each frequency interval (i.e., the emission becomes coherent).

As in all lasers, to obtain enough photons the junction has to be oriented in the direction of an optical cavity (two mirrors in front of each other, as with the ruby laser). If high-quality emission is required, these mirrors are external, but at the beginning of the laser era, cleavage planes were used as mirrors. Also, early on, the lasers worked in the infrared, they were pulsed, and they worked at liquid nitrogen temperature. Nowadays they work at room temperature, emit continuously in the visible region, and are used in compact disk and CDROM players.[1]

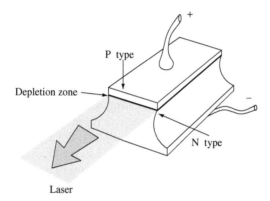

Figure 8.16 Sketch of a GaAs semiconductor laser.

These lasers emit multimode coherent radiation; that is, they emit stimulated radiation of varied frequencies. To filter them, the scheme of Fabry-Pérot laser mirrors can be substituted by distributed Bragg reflectors (DBRs) leading to a single-mode *distributed feedback semiconductor laser*. The spectral output band is typically 3 MHz wide, however. To fix this, one can build an extended cavity (at the output), where the beam is directed to a feedback loop in order to achieve a linewidth 12 times narrower.

Different technologies can control the distribution and thickness of different layers of semiconductors, and consequently semiconductor diode lasers with different properties can be obtained that make the laser emission more efficient. By reducing the active region thickness to 4–20 nm, the density of electrons and holes is modified and recombination becomes more efficient, lowering the laser threshold. These are known as *quantum well lasers* (either single or multiple). Further improvements have led to the fabrication of a new type of diode called a *vertical cavity surface-emitting laser* (VCSEL). The words "vertical emitting surface" mean in this context that laser emission is produced perpendicular to the wafer. Several significant advantages of this kind of diode (smaller size, narrower linewidth, circular beam, lower power consumption, higher efficiency) make them among the hottest areas in optical communications research today.

---

[1]Compact Disk Read Only Memory.

8.6 EFFECTS OF THE CRYSTALLINE STRUCTURE ON THE REFRACTIVE INDEX

Induced Anisotropies

The electromagnetic model that helped us to estimate the refractive index in a dielectric allows one to analyze properties linked directly to the crystalline structure. In the initial estimation we assumed that the materials are homogeneous and isotropic; that is, the properties and the optical parameters are the same in any direction in space and for any piece of material. Considering the refractive index, this implies that the propagation speed of an electromagnetic wave in this medium will be independent of the direction of propagation.

Some properties are linked in a more direct way to the material's crystalline structure than to its chemical composition; for example, the *birefringence* that appears in crystalline quartz does not appear in amorphous silica. Materials that naturally have these anisotropies are topics of optics and crystallography.

Years ago, during the development of lasers, the acquisition of crystals became quite important in developing new materials. Since their properties of anisotropy can be controlled by modifying the crystalline lattice with electrical (Pockels effect) or magnetic fields (Faraday effect) or mechanical stresses (photoelasticity), these materials are widely used.

Applying an electrical field to some materials enables us to modify any parameter of an electromagnetic wave such as its amplitude, polarization, frequency, or phase. This permits us to modulate radiation with information at lower frequencies (audio, video, or microwaves). The impact of such materials as KDP (KH_2PO_4), LN ($LiNbO_3$), ADP ($NH_4H_2PO_4$), and KTP ($KTiOPO_4$) has improved communications technology by moving into the optical range signals that before were, at a maximum, in the microwave region. For example, the Faraday effect allows one to build optical diodes and to observe the optical Hall effect in amorphous, vitreous, or granular materials.

The simultaneous development of semiconductor lasers, fast optical detectors, and optical fibers has given rise to optoelectronics, a new specialty of materials science that is growing fast, given the technology based on silicon, which is the earth's second most abundant element thus lowering the cost.

Nonlinear Phenomena

The ability to obtain laser radiation sources capable of generating light pulses of about 10^9 W has introduced the possibility of observing phenomena unknown previously.

Up to this point we have examined the optical characteristics of materials using the refractive index, which was taken as independent of the light intensity. When we study nonlinear phenomena in optics, the refractive index becomes dependent on the light intensity. The electromagnetic fields associated with the propagating wave can create these effects when their values are high enough. Sometimes these effects are known as quasistatic effects (as in the Kerr effect, which induces a birefringence proportional to the squared external electrical field). The appearance of such effects caused by the fields associated with the light intensity itself is recent, and is possible only because of high-power lasers. Before then, the fields associated with available light sources were never powerful enough to observe these phenomena. The phenomena include harmonic generation (figure 8.17), self-focusing, optical rectification, and frequency mixing.

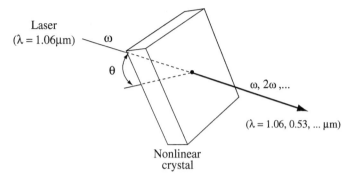

Figure 8.17 Harmonic generation by high-power light pulses.

As an example a short pulse in a neodymium-YAG laser of enough power, propagating in a nonlinear crystal, transforms part of the radiation (of infrared wavelength 1.06 μm) into radiation of green color (0.53 μm) that continues coherently and in the same direction. The applications of these phenomena in technology have hardly started yet.

Nonlinear crystals such as the periodically poled KTP (PP-KTP) suffer a quasi-phase-matched interaction with light. Thus, ferroelectric domains, periodically oriented or poled, induce a nonlinear interaction. This nonlinear interaction is quite useful since much less light is needed; additionally, the lattice's parameters can be changed in such a way that the nonlinear interaction is tunable. These systems are parametric optical oscillators.

PROBLEMS

8.1. Estimate the absorption coefficient of a metallic sheet, 50 nm thick, given that it reflects 40% and transmits 20% of the incident light.

8.2. Snell's law indicates that the refractive index multiplied by the sine of the angle formed by a luminous ray with the surface of separation between two media is constant when the ray goes through this surface of separation. In what situation is there no transmitted ray?

8.3. Optical fibers combine good transparency with the total reflection of the preceding problem. Discuss the function of optical fibers.

8.4. The simplest optical fiber can be regarded as a transparent cylindrical wire with a refractive index n placed in vacuum. What is the minimum value of n for which all the rays that penetrate the fiber experiment total reflection?

8.5. Discuss the existence of zones of anomalous dispersion.

8.6. Why are the Maxwell relation and the refractive index not the same (see table 8.2)?

8.7. On what magnitudes does the refractive index depend?

8.8. Calculate which wavelength causes a diamond to become opaque if its band gap is 5.33 eV.

8.9. One LED emits visible radiation with a visible wavelength of 0.52 μm. What is the energy level from which electrons relax to the valence band?

8.10. Can a laser work in a two-level system? Why?

8.11. Why is a tungsten bulb's light not coherent?

Chapter Nine

Noncrystalline Materials

The concept of the crystal introduced in chapter 2 includes materials whose long-range order has discrete translational invariance in the three spatial coordinates. In this chapter we look at a range of materials, from those without order (amorphous materials), to those with long-range order but without periodicity, and to those having short-range order comparable to interatomic distances (vitreous or glassy materials).

An ideal monocrystal has an ordered range equal to the sample size. If this solid has imperfections, the ordered range narrows by different mechanisms depending on the defects. For example, if the defects are planar, crystalline grains transform the sample into a polycrystalline one, whose mean grain size determines the order's range. If the defects are pointlike or are dislocations they shorten the correlation length, and the order's range changes accordingly.

All these materials display properties that are independent of the smooth transition from a perfectly ordered and periodic material (ideal crystal) to a disordered material. The range of the order is a continuum magnitude—which corresponds to a coherent length in other physicochemical systems—such that some properties sharply change at a fixed order while others vary smoothly from being perfectly ordered to being disordered (figure 9.1).

A crystal can transform into an amorphous or vitreous material in several ways that allow structurally different samples to coexist with the same order range in the same material. This transformation is called, in samples without long-range order, polyamorphism.

The translational invariance of a monocrystal is reflected in its Fourier spectrum. It leads to peaks defined by three integer numbers (Miller indices). In a quasicrystal, translational invariance is lost and the peaks of its Fourier spectrum cannot be labeled by three integer numbers. Here the Fourier spectrum consists of a dense set of sharp peaks whose position is determined by the sum of integer multiples of two numbers that have an irrational ratio. That is, the positions of atoms, ions, and molecules are determined by the sum of two incommensurate periodic functions that exemplify the property of quasiperiodicity. The canonical way to label a Fourier peak is to specify a number of integers greater than the space dimension. Thus, to describe some quasicrystals one must use six Miller indices, three more than a monocrystal needs. The geometrical origin of quasicrystals is that some *unit cells* do not match periodically. An example of this is pentagonal unit cells in two dimensions. Here one or more cells (called tiles), complement each other and matching in a nonperiodic way, can fill the entire space without intersections—just as classical tilings of a plane do. The plane's classical tilings can be visualized as the abstraction of a tiled floor. From here on, we generalize this concept to three-dimensional space.

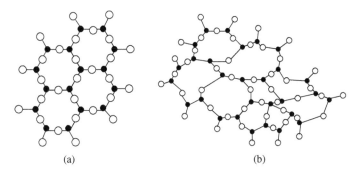

Figure 9.1 Sketch representing (a) crystalline and (b) vitreous SiO_2.

9.1 QUASICRYSTALS

Quasicrystals have three ingredients. First, they have quasiperiodic translational order; that is, the quasicrystal density is a quasiperiodic function. Second, there is a minimum separation between atoms, which is different from a superposition of two incommensurate periodic lattices. Finally, they have orientational order; in essence, the bond angles between neighboring atoms have long-range correlation, oriented along the axis defining this order.

Introduction

As quasicrystals can be described geometrically by plane or space tilings, it is key to know some of their relevant properties, for example, nonperiodic tilings.[1] Tilings might be only nonperiodic (figure 9.2); with these tiles it is impossible to build a periodic tiling. An example of this tile set is the Penrose tile set.

With a tile set, the tiling is not uniquely determined. This is an important point, one that calls for matching rules between tiles to create quasiperiodic structures. The matching rule determines, for example, the relative orientation of two tiles when they join together (figure 9.3)

Matching rules are not strict and have a subtle action: Given a set of tiles, many clusters can be built by following the matching rules. Yet the number of valid clusters that obey the rules decreases when the clusters grow. Thus, it is difficult to build a perfect quasicrystalline lattice by adding tiles one after the other; usually the additions block the way even when all matching rules are followed. So the matching rules do not prevent the appearance of defects. In an ideal sense, one could build a quasicrystalline lattice without defects:

- Through deflation rules that substitute a tile with a cluster.

- Through the projection or cut of a periodic structure in hyperspace. This is possible because orientation symmetry is not compatible with spatial groups in physical space. These forbidden point symmetries can be obtained through periodic tiling by increasing the space dimension. It is then possible to describe the reciprocal lattice of a quasicrystal as the projection of n (greater than the space dimension) needed vectors. For instance, in a quasicrystal with icosahedral symmetry, six vectors are

---

[1] They are ones where tile disposition lacks periodicity.

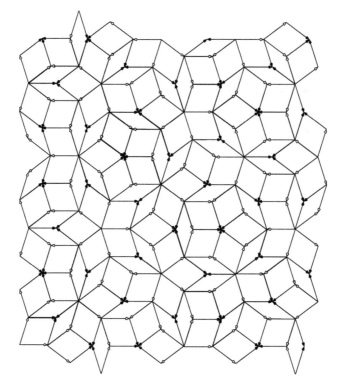

Figure 9.2 Nonperiodic tiling of the plane through the Penrose tile set.

Figure 9.3 Penrose tile set and its matching rules. The tiles must join in such a way that the arrow
types and directions agree.

necessary to describe the reciprocal lattice through integer combinations. These six
vectors actually result in a projection to a three-dimensional subspace with an incom-
mensurable slope. In this way vectors of the six-dimensional reciprocal lattice are of
the following type:

$$n_1\vec{b}_1^* + n_2\vec{b}_2^* + n_3\vec{b}_3^* + n_4\vec{b}_4^* + n_5\vec{b}_5^* + n_6\vec{b}_6^*.$$

Once the projection is done, we reach three-dimensional space and obtain six Miller
indices (h, h', k, k', l, l'). This description gives good results, but one should be
careful not to extrapolate to consequences that might seem logical—for example,
that the degrees of freedom of a quasicrystal are double those of a three-dimensional
crystal (figure 9.4).

• Through quasilocal growth as explained below.

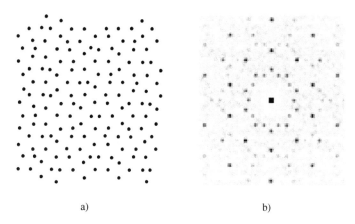

a)   b)

Figure 9.4 Monatomic lattice of Penrose tiles (a) and its Fourier spectrum (b).

Preparation of Quasicrystals

The first quasicrystals studied were low-quality AlMn quasicrystals, prepared by rapid quench procedures. Later, perfect icosahedra were obtained in AlCuFe and AlPdMn alloys. It is also possible to bring about reversible transitions from quasicrystalline to crystalline phases, which experimentally proves the difference between true icosahedral structures and microcrystalline pseudoicosahedral structures.

The most relevant preparation methods for quasicrystals are listed below.

Melt-spinning technique. A similar procedure is used for preparing metallic glasses and amorphous ferromagnets (figure 9.5) (sections 9.3 and 9.4), where the necessary cooling rates range from 10^5 to 10^9 K $\cdot$ s^{-1}, to avoid the formation of high-temperature equilibrium phases. Here melted alloys are poured onto a rotating disk that permits quenching rates to be as high as 10^6 K $\cdot$ s^{-1}. If we vary the disk's rotation speed, the quench rate is modified. The sample is obtained as a ribbon of several micrometers thickness and several millimeters width. These ribbons contain monoquasicrystals of about 1 μm in size. These sizes forbid characterization by X-radiation diffraction or neutron scattering. It is possible to use electron diffraction, however.

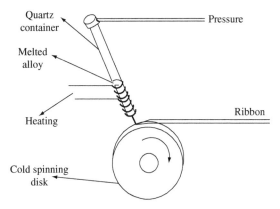

Figure 9.5 Sketch of the melt-spinning technique.

It is also possible to obtain monoquasicrystals, but with some trouble and scant repro-ducibility because of weak control over the procedure's parameters. The $Al_{80}Mn_{20}$ qua-sicrystal is obtained coexisting with other crystalline phases. Yet if a small quantity of Si (~5%) is added, it forms a unique quasicrystalline phase.

All useful methods for preparing metastable alloys and glasses may be applied to the preparation of quasicrystals. These methods are based on the production of disorder at the atomic level, generally by solid-state reactions, and can be summarized as follows:

Multilayer deposition technique. After a multilayer is achieved that has adequate com-position, it is subjected to bombardment by high-energy inert atoms.

Direct implantation, for instance, Mn atoms in an oriented Al matrix.

Mechanical alloying.

Vaporization technique. This is not a solid-state reaction; rather, it consists of quenching a fog of alloy droplets. The resulting smoke contains irregular shape particles with sizes in the 500–3000 Å range.

Thin-layer melting by laser or e^-. Neither is a solid-state reaction. The thin layer is locally melted by a laser or by high-energy electrons. The quench comes from radiative effects and from the thermal capacity of either the whole sample or the substrate or both.

Slow cooling of the melt. This technique is useful only when the quasicrystalline phase is stable over a range of relevant thermodynamic variables. It is not based on local disorder production, but on the contrary process. It allows large grains to grow into a structure that can be studied by X-ray diffraction or neutron diffraction. The first material obtained in this category was Al_6Li_3Cu (the icosahedral phase). For congruent growth—that is, to have the same composition in the solid as in the melt—it is necessary that the quasicrystal is prepared in a geode mode.

Icosahedral stable phases occur in GaMgZn, AlPdMn, and AlCu(Fe,Os,Ru) (figure 9.6), and decagonal phases in the systems AlCuCo and AlCuNi.

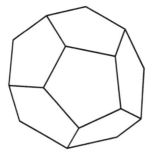

Figure 9.6 Outline of the dodecahedral monoquasicrystal of an icosahedral phase in an alloy of the system AlCuFe.

In each instance we should distinguish between true quasicrystalline behavior and para-sitic ordered polycrystalline samples which lead to diffraction patterns similar to the ones in quasicrystals.[2]

Referring to the models, atoms should be assigned to each tile vertex in a tile set with matching rules. This is called atomic decoration (figure 9.7). Sometimes, this leads to two or more equivalent orientations for the atomic decoration but not for the matching rules. This

---

[2]Note that every well-behaved function can be approximated to a sum of periodical functions.

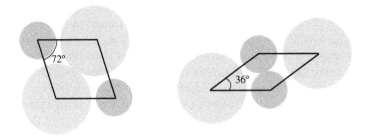

Figure 9.7 Example of atomic decoration of the Penrose tile set (figure 9.3).

equivalence creates the local isomorphism classes of a quasicrystal and does not indicate true lattice defects.

As we noted above, the classical matching rules (which reveal the relative orientations between two compatible tiles by local matching laws at their edges or faces) are not sufficient to determine a quasicrystalline structure and avoid defects. If we introduce new matching rules related to the vertices, however, we can grow locally almost perfect quasicrystals. With the new matching rules the ambiguities in building quasicrystals diminish.

Quasicrystal growth is described by phenomenological theories like the Landau phase transition theory, based on symmetry and mean-field considerations. Thus, Landau theory leads to an equilibrium transition between disordered states and icosahedral-type states. A more specific theory describing quasicrystal growth is the random tiling model, which can explain the growth of quasicrystals with many defects and disordered quasicrystals. Also, we have theories describing growth, from structures already formed, by relaxation processes.

Quasicrystal Types

The most relevant quasicrystal types are

- Unidimensional phases (or Fibonacci phases).

- Octagonal phases.

- Decagonal phases.

- Dodecagonal phases.

- Icosahedral phases. They are the most common, having various subtypes:

  - AlMn-like phases. They are metastable and correspond to a simple cubic lattice in six dimensions. They lead to vitreous materials.

  - Phases of Bergmann-like clusters with local atomic order. The mean structure is also a six-dimensional simple cubic one. They originate from random tilings along with entropy arguments. They have a small quasicrystalline fraction, a good example being the AlLiCu-like alloys. An exception is the AlMgLi-like alloy corresponding to a six-dimensional fcc lattice.

    – Phases of Mackay-like clusters with atomic local order. These quasicrystals are nearly perfect, with Bragg peaks without diffuse dispersion. They are six-dimensional fcc lattices. They may form true monoquasicrystals with new properties different from those of vitreous and crystalline materials. Some interesting examples of this type of icosahedral phase are $Al_{0.62}Cu_{0.255}Fe_{0.125}$, $Al_{0.695}Pd_{0.255}Mn_{0.08}$, $Al_{0.64}Cu_{0.22}Ru_{0.14}$, and $Al_{0.71}Pd_{0.2}Re_{0.09}$. In some, dynamic diffraction (the Borrmann effect) has been observed, where second-order coherent interaction comes into play and proves that there is a very long-range order.

Quasicrystal Characterization

The characterization of quasicrystals can be performed by different kinds of diffraction. In principle, the diffraction in monoquasicrystals is the only way to characterize appropriately quasicrystalline samples.

X-ray Diffraction

Electromagnetic radiation interacts with an atom's electrons. These electrons are accelerated by the $\vec{E}$ field from X radiation such that secondary X rays are emitted, of equal frequency and phase, but with an amplitude[3] proportional to the atom's number of electrons. This method is highly sensitive to the incoming angle. The wavelengths range from 0.7 to 2 Å. Interaction with the samples is strong; they are required to be larger than 0.1 mm. Also, it is possible to do powder X-ray diffraction.

Neutron Diffraction

Neutrons interact with the atomic nucleus. For an instant the neutrons are captured and form a nucleus that reemits the neutron in phase (positive) or out of phase (negative), according to the dispersion length. Two different samples can lead to two similar X-ray diffraction spectra but different neutron diffraction spectra. For instance, ^{6}Li (with a coherent dispersion length of $+0.2 \times 10^{-12}$ cm) and ^{7}Li (with a coherent dispersion length of -0.222×10^{-12} cm) lead to quite different spectra. This enables us to distinguish two samples containing different isotopes, and it allows us to prove that if the two spectra are the same the order is not related to the isotopes. The interaction is weak; consequently, the samples should be as big as 1 cm. The wavelengths of neutrons range from 0.7 to 3.5 Å (although with cold neutrons one can attain up to several tens of angstroms). Again, it is possible to do powder neutron diffraction.

Electron Diffraction

Electrons are strongly scattered by the atomic potential. They are strongly absorbed; hence, only thin films can be studied. The electron wavelengths are of the order of 2.59×10^{-2} Å. The most useful characterization is by selected area electron diffraction (SAED), which is done for samples several hundred angstroms thick, with areas of $0.1 \times 0.1 \mu m^2$, and small quasicrystalline grains. We can also use high-resolution electron microscopy (HREM), whose resolution is around a few angstroms. The problem is to link the real structure with

---

[3]Called "atomic form factor."

the image obtained because of dynamic effects and parameters related to the measurement apparatus.

Quasicrystal Properties

When considering the mechanical and thermal properties, remember that disorder expands vibrations in reciprocal space and thus confines them in direct space. This phenomenon is called weak localization.

In disordered media, there are no phononic modes (neither optical nor acoustic); therefore, the vibrations cannot be described correctly by the dispersion law $\omega = f(\vec{q})$. The excitations have a shorter mean lifetime and are confined. The related state density is controlled by short-range order and cluster formation effects.

In fractal media (remember that fractals are self-similar, and this self-similarity leads to the specific properties of this kind of medium) the vibratory modes are called fractons. These have two different regimes of behavior, the first corresponding to wavelengths smaller than the characteristic length of the fractal, ξ, and characterized by localized vibrations, the other one corresponding to $\lambda > \xi$ (low-energy modes) and characterized by the existence of dissipative traveling modes.

There is a partial analogy between phonons and other collective modes like phasons. While the former are traveling modes[4] represented in direct space, the latter are diffusive and are represented in reciprocal space. This implies that there are unit-cell readjustments in a diffusive-like relaxation.

The number of optoacoustic branches in a periodically modulated lattice is equal to the ratio of the two periodicities. But if the lattice is quasiperiodic the number of branches is equal to the number of Fourier peaks; hence the branches form a dense set.

When a strain field acts on a lattice and the field's phase or amplitude changes over time, collective modes appear; they are named phasons and amplitudons, respectively. Both play a major role in defective quasicrystal structures.

Experiments (in icosahedral AlPdMn) lead to the following results:

- The dispersion relations show two pseudobranches (transverse and longitudinal).

- For long wavelengths ($\lambda > 3.33$ Å or $\omega < 1.4$ THz or $E < 6$ meV) isotropic linear acoustic modes appear.

- For $1.67 < \lambda < 3.33$ Å or $2.8 > \omega > 1.4$ THz or $12 > E > 6$ meV, the excitation broadens progressively and the linewidth for a given $\vec{q}$ reaches 1 THz (4 meV). The dispersion relation is nonlinear but isotropic.

- For $\lambda < 1.67$ Å or $\omega > 3$ THz or $E > 12$ meV, the vibration is not dispersive and the branches are flat.

In short, the global vibratory behavior in quasicrystals has two contributions, phononlike and phasonlike. The distinctive behavior at different wavelengths comes from deflation growth laws that have a pronounced self-similar feature, like the one that occurs in fractal lattices.

The dynamic behavior of topological defects[5] in quasicrystals is different from the behavior of those in crystals because of the higher-dimensional space. Thus, although there

---

[4]Consequently, they couple to $\vec{P}$.

[5]That is, defects with a non-zero Burgers vector.

were defects, the quasicrystal would be brittle and would lack appreciable plasticity. The
dislocations cannot move easily because they involve phasons and diffusive time scales.
The Burgers vector is defined in hyperspace as having phononlike and phasonlike aspects.
If we cut a quasicrystal and paste the new edges, we do not obtain a dislocation but a
stacking fault, which is a planar defect.

The electrical properties, some of which are not yet thoroughly explained, can be summarized as follows:

- The resistivity at low temperatures is low. Yet it is two orders of magnitude greater
  than that of amorphous metals and four greater than that of crystalline metals (figure 9.8). It reveals a metal-insulator transition (figure 9.9).

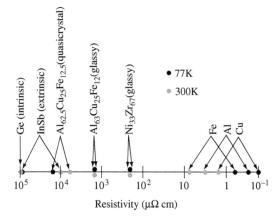

Figure 9.8 Resistivity of a quasicrystal compared to that of semiconductors, metallic glasses, and
crystalline metals.

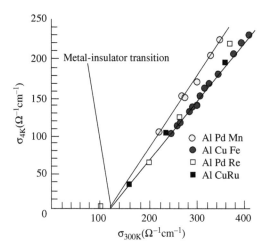

Figure 9.9 Relation between the conductivity at 4 and 300 K for some icosahedral quasicrystals.

- The density of states at the Fermi level is similar to that of a semimetal (10^{20} cm^{-3})
  (figure 9.10).

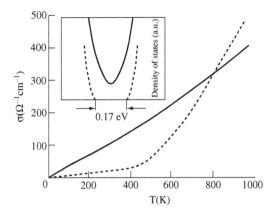

Figure 9.10 Electrical conductivity of the $Al_{70}Pd_{20}Re_{10}$ quasicrystal (solid line) and of the Al_2Ru semiconductor (dashed line) as a function of temperature. The effect of the low electronic density of states below the forbidden zone of the semiconductor can be seen.

- If a good conductor is added at the proportion of 1% by volume, this short-circuits the sample and radically decreases its resistivity.

- The ratio between the resistivities at 4.2 and 290 K is around 10, while in amorphous metals it is of the order of unity and in crystalline metals it is smaller than 0.1 (figure 9.8).

- The electron mean free path is 20–30 Å and therefore much smaller than the Fermi length.

- The optical conductivity cannot be explained by Drude-like models.

- The resistivity is sensitive to symmetry breaks related to the appearance of imperfections, in the opposite way. That is, imperfections make the resistivity decrease considerably (figures 9.11 and 9.12).

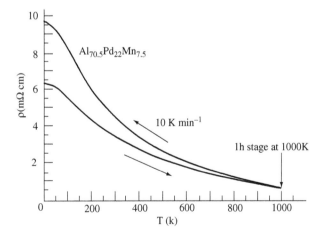

Figure 9.11 Hysteresis in the resistivity of the $Al_{70.5}Pd_{22}Mn_{7.5}$ quasicrystal. This proves that resistivity increases when defects decrease after undergoing a thermal treatment.

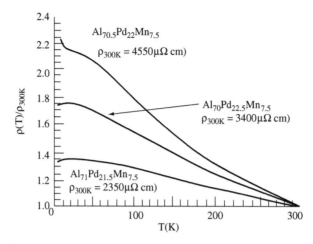

Figure 9.12 Reduced resistivities as a function of temperature in different quasicrystals of the AlPdMn system. Note that resistivity increases when the quasicrystal becomes more perfect.

Other physicochemical properties, not yet well established, are the following:

- The specific heat also shows a small density of states at the Fermi level, although some measurements may be screened by imperfections.

- In the most perfect icosahedra the Hall coefficients are slightly negative, changing sign when temperatures exceed a certain amount. Also, these measurements are quite sensitive to imperfections (figure 9.13).

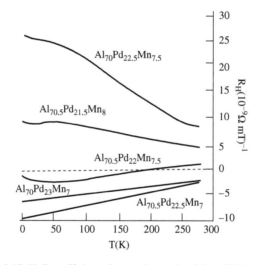

Figure 9.13 Hall coefficients for quasicrystals of the AlPdMn system.

- The thermoelectric behavior is different from that of metallic glasses, being strongly nonlinear and also showing a sign change (figure 9.14).

- Photoemission and absorption are coherent with $g(E_F)$.

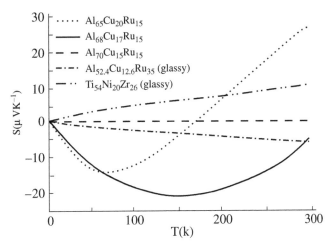

Figure 9.14 Thermoelectric power as a function of temperature for quasicrystals in the AlCuRu system and for metallic glasses. Nonlinearities and sign changes can be detected in quasicrystals.

- At the end of the twentieth century icosahedral quasicrystals with long-range magnetic order were discovered. The expected order corresponds to that of antiferromagnetic materials.

Decagonal quasicrystals (presenting two-dimensional quasicrystalline order and periodic order in the other dimension) have intermediate behavior due to the combination of the two ingredients, hence leading to anisotropy.

9.2 GLASSES AND GLASSY MATERIALS

Vitreous materials present short-range order. Glasses should be distinguished from nano-crystalline materials that have nanometric grains, from polycrystalline materials (obtainable from nanocrystal annealing) with microscopic or macroscopic grains, and from quasicrystals that were analyzed in section 9.1. All of them have short- or long-range order. Vitreous materials should also be distinguished from other disordered materials such as the generic amorphous materials. Vitreous materials differ from amorphous materials because they continuously form from the liquid state by supercooling to lower temperature than the solidification temperature; so the structure coming from short-range order or from statistical properties remains. Nevertheless, other amorphous phases can exist and may even coexist. This last phenomenon is called polyamorphism. As an example, Si in the liquid state at atmospheric pressure is dense and behaves like a metal. But amorphous Si is created by thin-layer deposition and presents covalent-like bonds that are tetrahedrally coordinated, as is the crystalline solid. Vitreous Si has properties more like those of liquids than crystals.

Supercooling must happen at cooling rates greater than those that withstand crystalliza-tion. The temperature corresponding to the glass transition must be derived using some convention because the properties do not change abruptly at the transition. Usually the *glass transition temperature* is that where a large drop in specific heat at constant pressure (with values ranging from 40% to 100% of the vibratory specific heat at constant pressure) creates a transition from ergodic phases, where the system goes through all the states in the phase space, to nonergodic phases (figure 9.15). But this is a purely dynamic transition; there is no equivalent thermodynamic phase transition. This affirmation has been proved by numerical simulations [24].

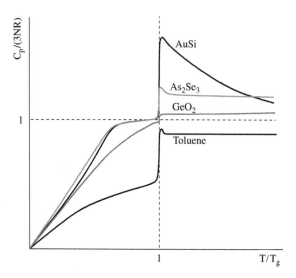

Figure 9.15 Glass transitions for molecular (toluene), metallic (Au-Si alloy), covalent (As$_2$Se$_3$), and open-network-like (GeO$_2$) materials.

The glass transition defines a range of temperatures where the system goes out of equilibrium. This range is necessary to change the mean relaxation time by 2–3 orders of magnitude (from 0.1 to 100 s). The slowing of the relaxation time is accompanied by a diffusive slowdown because there is less kinetic energy in the material.

Usually,[6] at the glass transition temperature the material's *viscosity* rises to 10^{13} P (10^{12} kg/m · s). The viscosity makes a material either a solid or a liquid; that is, the viscosity sets the material shape or frees it because it is connected with the usual time for the material to creep or lose its shape. Few liquids can be cooled fast enough to form a solid before crystallization starts. In this sense, the liquid's *mobility* is high enough that the ordering of the liquid-solid transition usually happens long before the viscosity rises to 10^{13} P. For these reasons, the glass transition temperature is usually the one where high viscosity values are attained.

On the whole, we need high cooling rates (figure 9.16). For instance, liquid water at 0°C has a viscosity of 1.8 cP; thus, the molecules are ordered fast enough to crystallize before the viscosity rises above the glass transition point. There are some good glass-forming ceramic materials like B$_2$O$_3$ (the main ingredient[7] of Pyrex® glass) or SiO$_2$, the main ingredient of common glass.

Reordering that happens during a sample's crystallization may be avoided by making the melt more complex. This complexity can usually be attained by adding new components to the melt that create new degrees of freedom, which muddles the system and makes crystallization more difficult.

---

[6]Except in brittle molecular liquids.

[7]To crystallize, it is necessary to apply pressure.

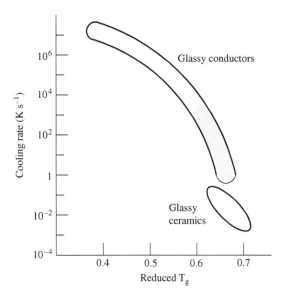

Figure 9.16 Necessary cooling rates for different kinds of glassy materials as a function of the ratio of the glass transition temperature to the fusion temperature. Low cooling rates correspond to glasses and glassy ceramics.

A fluid's viscosity is related to how fast the molecules go to equilibrium in a small-scale reordering. This reordering may take place either by vacancy diffusion or by interchange between two neighboring molecules; in both instances an electromagnetic energy barrier opposes the reordering. If the barrier is much higher than the thermal energy, the viscosity will be high. If not it will be low. The order of magnitude of the minimal viscosity in a generic fluid can be estimated by considering why the fluid molecules interact more than they do in a gas with the same physicochemical parameters (intermolecular distance, density, etc.). Consequently, the fluid viscosity will be greater than that of the gas; its viscosity is easily calculated, obtaining a typical value of 0.3 cP. This yields the barrier potential's order of magnitude if we fix the value of the glass transition temperature. For example, if we want the glass transition to occur at 2000 K, then the energy barrier will be approximately 6 eV. In general, the higher the energy barrier, the more vitreous the material. So the energetic barrier's order of magnitude in glass-forming materials is greater than 1 eV and of the order of 0.01 eV in materials that are bad at glass forming.

In the same way, activation energy barriers for magnetic materials may lead to crystalline materials with magnetic disorder.

One application of glasses comes from the ease of molding them at temperatures high enough that their viscosities are much lower than at room temperature. For instance, vitreous materials can be molded at viscosities ranging from 10^3 to 10^6 P, which corresponds to a temperature of about 1000°C for common glass ($SiO_2 + 25\%Na_2O$).

Vitreous materials exhibit properties depending on the overcooled liquid forming them. Among these, relaxation effects make the vitreous materials creep or flow, although with very large times, as common fluids do. Also, devitrification may be brought about by raising the temperature to values between the glass transition temperature and the fusion temperature. Devitrification happens via nucleation and growth of crystalline grains in the same way as solidification (crystallization), creating a non-glass-forming liquid.

It is useful to visualize glassy materials in topographical models whose total potential energy Φ (figure 9.17) is plotted against the $3N$ coordinates corresponding to atoms, ions, or molecules. As is well known, the minima of the hypersurface determined by Φ, indicate the system's stationary behaviors. It is clear that at $T = 0$ K the system goes to a minimal energy state. Usually, this state is a crystalline or quasicrystalline phase. Local minima above the ground state are amorphous packings visited by the stable liquid phase at temperatures greater than the fusion temperature.

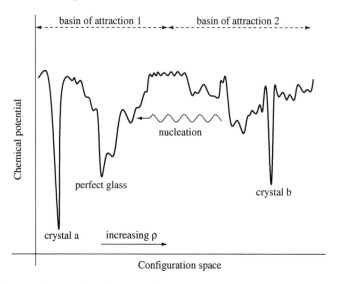

Figure 9.17 Diagram of a transition between different vitreous states through nucleation of one phase into the other, various crystalline phases, and an ideal vitreous phase. Several basins of attraction also appear. Note that the figure is extremely schematic because the configuration space is $3N - d$. This could explain some hidden paths between transitions. Moreover, the crystalline permutations are not shown.

Topographical models permit one to identify different kinds of relaxation (figure 9.17). The first, called α relaxations, are related to transitions between different basins of attraction and exist for all glassy materials, mainly because they have temperatures greater than T_g, leading to creep flow. The second, called β relaxations, are related to transitions inside a basin of attraction that have temperatures lower than T_g. These relaxations are characteristic of brittle liquids without viscous behavior corresponding to an Arrhenius law.

Also, in good glass-forming brittle liquids near T_g, the hydrodynamic diffusive constant is two orders of magnitude greater than expected from the viscosity at that temperature calculated with the Stokes-Einstein relation. This is due to the greater diffusion in regions locally fluidized by β transitions. Liquid silica reveals a brittle-to-strong transition leading to polyamorphism. This is because it is a good glass-forming material [25].

The existence of ideal vitreous states is not yet established. Such a state would be in the equilibrium phase (or that of minimal energy). They have been theorized at temperatures lower than the Kauzmann temperature, but they have not been discovered in experiments.

Many techniques are common for preparing glassy materials. They are conventional casting, vapor condensation (the main mirror of Mount Palomar Observatory was created in this way), heavy-particle bombardment of the crystal, and organic compound hydrolysis, followed by drying, vapor phase reactions (with pure reagents), and ending with condensation (a normal method for fabricating optical fibers). The most innovative techniques are those that create metallic glasses, amorphous ferromagnets, and quasicrystals (section 9.1).

9.3 METALLIC GLASSES

An amorphous material with vitreous properties is frequently made by injecting an atom jet into a low-temperature substrate. This is useful for creating amorphous alloys at industrial quantities. The most important example is the melt-spinning technique (section 9.1). This technique means that it is only possible to obtain metallic glasses or amorphous alloys in the shape of sheets or ribbons. Other methods of preparation are similar to the ones used in quasicrystals.

The main property of metallic glasses is work softening, opposite to what occurs in crystalline metals. A small plastic strain leads to a brittle fracture. Crystalline displacements are not possible, which implies high stresses generated by plastic strains and, therefore, fracture at low plastic strains. This wear during use is advantageous in metallic glass applications. Soft magnetic properties characterize this type of material as we find in section 9.4. Yet sometimes these lead to problems like strong magnetostriction because of the composition and low saturation magnetization.

It is possible to cause devitrification as in all glassy materials. Devitrification improves some properties by increasing ductility, by raising the flow stress, and by fracture resistance. The coercive force declines and the saturation magnetization climbs. In type II superconductors devitrification raises the critical current. All these improvements happen because devitrification makes a locally monocrystalline (or polycrystalline) material that is nearly monodisperse.

9.4 AMORPHOUS FERROMAGNETS

Amorphous ferromagnets are of interest because the disorder results in nearly isotropic properties and hence the magnetocrystalline anisotropy is almost zero. These materials also have low coercive forces, high permeabilities, and small losses in hysteresis loops. Because it is an amorphous alloy, there is no positional order for the chemical species; the electrical resistivity is higher than the resistivity of its crystalline ordered counterpart. All these properties give amorphous ferromagnets applications as soft magnetic materials.

They are prepared by the melt-spinning technique which quickly cools (quenches) a melted alloy jet on a rotating cold disk (see section 9.1).

Transition metal–semimetal alloys (transition metal 80%, usually Fe, Co or Ni; semimetal B, C, Si, P, or Al) are intriguing. The semimetal lowers the fusion point and stabilizes the alloy's amorphous phase. For instance, $Fe_{80}B_{20}$ (Metglas 2605) has a glass transition temperature of $T_g = 441°C$, 1538°C being the fusion temperature of pure iron. The Curie temperature of this material is 674 K, the coercive force is 0.04 G, and it has a magnetic permeability of 3×10^5. Mumetal$^®$ ($Ni_{0.77}Fe_{0.14}Cu_{0.05}Mo_{0.04}$) has a high permeability that makes it useful for electromagnetic screening. In this kind of material the coercive force can be as low as 0.006 G.

Materials with high coercive forces can be obtained when a polycrystalline material is made (less angular velocity during its creation) with a grain size equal those of the magnetic domains. In this way, all the domains may become well aligned. For greater grain sizes the magnetization is balanced inside each grain; for smaller grain sizes (more disordered) the system is isotropic, and the grain limits have a significant demagnetizing influence. For example, in the metastable $Nd_{0.4}Fe_{0.6}$ alloy the coercive force can be as high as 7.5 kG. For the $Sm_{0.4}Fe_{0.6}$ alloy one can reach coercive forces as high as 24 kG (figure 9.18).

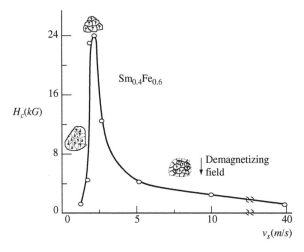

Figure 9.18 Coercive force at room temperature versus linear rotation velocity during fabrication. The relation of magnetic domains and crystalline grains is depicted.

9.5 AMORPHOUS SEMICONDUCTORS

Amorphous semiconductors are obtained by vaporizing or sputtering in thin films or sometimes in volume as regular glasses (i.e., by supercooling the melt).

Amorphous materials, specifically amorphous semiconductors, cannot be understood by applying the Bloch theorem because the crystal field is not periodic here. The main consequence is that the energy bands are not well defined and do not have the reciprocal lattice's periodicity, even though one can construct a band theory. In amorphous materials, the electronic states do not have well-defined values of $\vec{k}$; thus, the optical transitions have quite relaxed selection rules that receive contributions from the infrared and Raman modes to the spectrum.

A band structure does exist, but that is because the materials have similar quantum electronic configurations close to each other; this leads to splits in quantum levels owing to Pauli's exclusion principle. But here the splitting is inhomogeneous because of the diverse interatomic distances. By contrast, the disordered structure reduces the mean free path of electrons and holes until it reaches its characteristic length scale because they are scattered when they move.

Localized states can also occur in amorphous solids due to the inhomogeneity of the crystal field. These localized states remain close to the extrema of the energy bands; namely, they occur for low-energy electrons and holes which in few collisions may lose their kinetic energy. In all these states, conduction takes place as a thermal promotion. This leads to an anomalous Hall effect at low temperatures because the localized states, pinned to the inhomogeneities, do not conduct, and the Hall effect is not useful for determining the carrier concentrations.

Thus, amorphous semiconductors behave like intrinsic semiconductors, except that the Fermi level is pinned because of the inhomogeneities that generate localized states in the forbidden zone.

The main types of amorphous semiconductors are those with tetrahedral bonds, those formed with chalcogen multicomponent solids (with S, Se, or Te), and compounds of elements in group V of the periodic table. The first have similar properties to their crystalline counterparts, but the details come from inhomogeneities. Thus, if the inhomogeneities are compensated with hydrogen (10%), we obtain an amorphous material similar to a monocrystalline material (in its semiconductor-type properties) and much cheaper. This permits one to build photovoltaic cells with them.

Amorphous semiconductors are useful in xerography and in simple computation and memorization devices. These are applications where the chalcogen multicomponent solids with an amorphous-crystalline reversible transition are used, owing to their ability to memorize, when they form an anisotropic crystalline phase from an amorphous, homogeneous, and isotropic phase.

9.6 LOW-ENERGY EXCITATIONS IN AMORPHOUS SOLIDS

Specific Heat

The specific heat of a crystalline dielectric at low temperatures follows the Debye T^3 law because long-wavelength phonons become excited. We expect the same behavior in vitreous or amorphous materials. But experimentally they behave linearly at low temperatures, which leads to specific heats 1000 times what is expected when following the Debye law. These anomalous contributions are not fully understood and do not match the linear term of the electronic contribution to the specific heat in metallic materials. We believe that this behavior exists because of excitation at two low-energy levels.

Thermal Conductivity

The thermal conductivity in vitreous and amorphous materials is tiny. At room temperature or greater, it is limited by the order's scale, because this determines the mean free path of the dominating thermal phonons. For low temperatures (< 1 K), the conductivity is limited by the phononic dispersion corresponding to the excitations for the specific heat.

PROBLEMS

9.1. How can you distinguish a crystalline material from a noncrystalline one?

9.2. Give examples of properties that change continuously with respect to the range of a material's order.

9.3. Give examples of properties that change discontinuously with respect to the range of a material's order.

9.4. In what ways is an amorphous material like and not like a polycrystalline one when the grain size approaches zero?

9.5. What is the characteristic time for a 10 cm edge cube of common glass to relax at room temperature?

9.6. Draw (in two dimensions) the structure of vitreous GeO_2 following figure 9.19, obtained by extended X-ray absorption fine-structure spectroscopy (EXAFS).

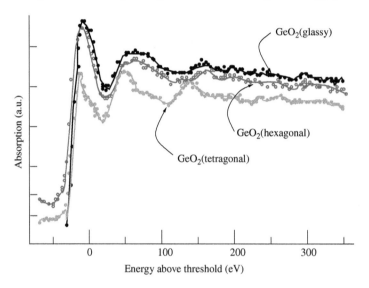

Figure 9.19 Absorption in materials made from GeO_2 as a function of the X-radiation energy relative to the threshold for the photoexcitation of an electron from a deep state (e.g., a K state) to an allowed band.

9.7. Usually, transport coefficients of noncrystalline materials are smaller than their crystalline counterparts. Why?

9.8. What kind of anisotropies are smaller in noncrystalline materials than in their crystalline counterparts?

9.9. What generic properties occur in a composite material that is amorphous in a crystalline matrix? Give arguments.

9.10. What generic properties occur in a composite material that is crystalline in an amorphous matrix? Give arguments.

Chapter Ten

Polymeric Materials

Among the materials of scientific and technological interest some are formed of gigantic molecules whose molecular weight exceeds $10^4 \, g \cdot mol^{-1}$ and which are called macromolecular substances. This kind of substance was used long ago without modifying its molecular structure as wood, rubber, cotton, or wool, either in the form of elaborate by-products like alkaloids or as natural dyes.

In the second half of the twentieth century two new families of macromolecular synthetic materials appeared. This altered the industrial world by invading almost all fields of applications. They were initially called *plastic materials* and *artificial rubbers* or elastomers. These materials are macromolecular substances in which the structural units are repeated regularly. For this reason they are called *polymeric materials*. The earliest example is Bakelite, a material produced by condensing formaldehyde and phenol, cresyl alcohol, or xylenol. Bakelite was discovered by Leo Hendrik Baekeland in 1907.

Industrial polymeric materials are formed by long chains of subunits or *mers*,[1] joined by covalent bonds. Due to their length and high molecular mass they are also called *high polymers*. They differ from other polymers and natural biopolymers, characteristic of organic substances like DNA, RNA, or proteins, by having been obtained through synthesis procedures (*synthetic polymers*). Those obtained by transformation, starting with natural products, created without an appreciable destruction of the original macromolecule, are called *semisynthetic polymers*. Biopolymers represent a key chapter in material science because of their blossoming number of applications. Examples of synthetic polymers obtained from low-molecular-mass substances are polyamide (PA, nylon) and polystyrene (PS). Examples of semisynthetic polymers are nitrocellulose, ebonite, and artificial silk.

For several decades the applications of high polymers focused on producing fibers for the textile industry, films, and coverings for construction of all kinds, and on low-cost substitutes for metals, which are thus saved for applications requiring more advanced technologies. But now polymeric materials have been incorporated by high-technology industries such as optic and electronic communications, computing, and the electronics industry, where the simple initial use as an insulating material has now been exceeded. This material has also established a major presence in substituting for aluminum and other structural materials in industries such as automobile manufacturing. This is due to the newly developed capacities to bear great mechanical stresses and high temperatures, and it also reduces the mass and the noise inside a vehicle.

From a structural viewpoint, high polymers form design materials with excellent prospects, since they admit design at different scales. At a microscopic scale one can design *custom molecules* to form the mers and assemble them with defined properties. At the mesoscopic scale we can manipulate either the structure or the orientation of macromolecules (conformation) to design custom materials. This controls how regions with different crys-

---

[1]From the Greek $\mu\epsilon\rho o\sigma$: portion. From the Latin *merus*: pure, simple.

talline composition are distributed. Some polymers can be blended to obtain the desired performance qualities. Finally, at macroscopic scales, we can use them as matrices of composites to achieve special or intermediate properties.

Classification

In high polymers, owing to their diversity and the extent of the areas under investigation, different classifications are chosen according to which aspects one wants to highlight.

1. Depending on their origin they can be classified as natural polymers, semisynthetic and synthetic, from nature, or from a synthesis of components without large modification. According to the formation reactions during the preparation they are normally classified as:

Polymerized,

Polycondensed, and

Polyadducted materials.

In polymerized materials, macromolecules are formed by joining nonsaturated monomer molecules via breaking the double bonds, without the simpler molecules drifting apart. A classic example is polyethylene (PE).

In polycondensed materials, at least two groups of monomers react and produce the bonds (bi-, tri-, or polyfunctional), leaving residues like H_2O, NH_3, HCl, and so forth. Some examples are nylon, cellulose, and proteins.

In polyadducted molecules, as in polycondensed, the reaction takes place between at least two polyfunctional groups, but simpler molecules do not separate. Examples are polyurethane and epoxy resins.

2. According to the shape of the macromolecule, the polymers are normally grouped into those formed by

Linear chains (parallel and in balls),

Ramified chains, and

Reticles (they include entangled chains).

3. Depending on the macromolecule's chemical composition, the polymers are classified as

Carbopolymers (they contain only carbon and hydrogen; C, H),

Carboxypolymers (they contain C, O, H),

Carboazopolymers (C, N, H, and sometimes O),

Carbotiopolymers (C, S, H, and sometimes O), and

Siloxypolymers (Si, O, H).

4. They also are grouped in relation to their physical properties or their *thermoelastic behavior* into

Elastomers,

Thermosets, and

Thermoplastics.

Elastomers have the elasticity of rubber. They are comprised of linear macromolecules joined transversely by bond bridges (vulcanization). During tensile or compression stresses, molecules are displaced as a group—which is what gives synthetic rubber its elasticity.

Thermoset plastics are materials that, because of heat or chemical agents (called *hardeners*), harden in an irreversible way. They are formed by reticulated molecules whose hardening action reticulates them even more. When heated up, they usually decompose before they melt. Examples of polycondensed and reticulated polyadducted polymers are the phenolic resins, the melamines, uric resins, and so on.

Thermoplastics are formed by linear macromolecules or those with little reticulation. If they are subjected to the action of heat they soften (or *plasticize*) in a reversible way, solidifying again when cooled. They melt without decomposing. Due to these properties they are appropriate either for molding when hot, or for injection and lamination, or for reduction into fibers.

In this chapter we pay special attention to synthetic polymeric materials and the possibilities that advances in synthesis and production offer in terms of properties that can be programmed.

10.1 MOLECULAR STRUCTURE

Molecular design has been the most important area that designers of materials have been able to influence. In analyzing the relation between the structure and properties of polymeric materials, we start by enquiring into the macromolecules that form them.

Macromolecules that form polymeric material are created by hundreds or thousands of subunits or mers, functional units that are repeated and kept tight by covalent bonds. Each unit comes from a main molecule that can be bonded to other similar units to form a chain. This chain may or may not have lateral ramifications and bond to others through cross-links.

The initial molecule is the monomer; it has no free bonds and can convert, under any action, one internal bond to a free link capable of being bonded to other units. This molecule includes one or several functional groups.

A homopolymer is a polymer formed by only one type of mer. In contrast, a heteropolymer or copolymer is created by more than one type of mer and permits many possibilities for continuously graded and optimized properties, depending on the application we desire. Among the most significant examples of heteropolymers are biopolymers like DNA, which has four types of monomers, and proteins, which can have up to 20 monomers.

Homopolymers

One simple and paradigmatic example of a homopolymer is polyethylene (PE):

$$\cdots - CH_2 - CH_2 - CH_2 - CH_2 - CH_2 - CH_2 - CH_2 - CH_2 - \cdots$$

obtained when heating ethylene in the presence of catalysts. This can be represented in this way:

$$n(CH_2 = CH_2) \longrightarrow (-CH_2 - CH_2-)_n.$$

In this macromolecule, the monomer is ethylene $CH_2=CH_2$. High polymers are macromolecules obtained from PE, whose molecular weight, with dispersal around its mean value, varies between 20,000 and 3,000,000. The degree of polymerization n indicates the average number of mers that form each molecule.

Many properties are modified once the number of mers rises. For example, the melting point and density change in such a way that they depend on the degree of polymerization and on the composition and chemical structure of the extremes of the chains. When the degree of polymerization increases, the influence of both ends of a chain becomes smaller and smaller; for large enough n, the properties are practically independent of the degree of polymerization. Figure 10.1 shows how the melting point varies as the molecular weight increases.

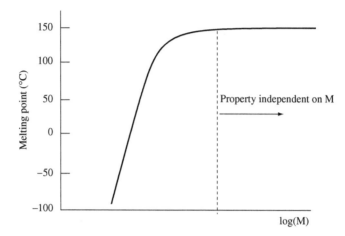

Figure 10.1 Melting point of a polymer as a function of the number of units of the chain.

When the number of units is small (under 200) the molecules are normally named oligomers, using the name "polymers" only when the degree of polymerization is higher and the properties are nearly independent of it.

Examples of changes occurring as the number of units in a macromolecule changes include ethylene, whose unit $-C_2H_4-$ turns into ethane ($n = 1$) and butane ($n = 2$), which are gases, n-octane ($n = 4$) which is a liquid, and the paraffins (from $n = 10$ to $n = 15$). If we continue increasing n we go through greases and waxes until we get to a degree of polymerization close to 200. We eventually reach high-density polyethylene (HDPE), a solid whose properties depend little on the degree of polymerization. HDPE is a thermoplastic that decomposes into subunits under the action of heat or solvents.

The skeleton that forms the chain is not always straight. The skeleton's main covalent bonds require the conservation of distances and angles. Even when it contains only C-C bonds, as shown in figure 10.2, the chain can be twisted in space because there is freedom to position the next carbon atom.

The spatial disposition of a segment of the chain of PE is displayed in figure 10.3. It is a sequence of carbon atoms bonded by strong covalent bonds. Each carbon atom is also bonded to two hydrogen atoms.

This spatial disposition—which is not unique, since it has rotary isomers—is responsible for the polymer chain's great flexibility. The equilibrium position of the atoms shown in the figure can be slightly modified by various causes. First, spatial changes in the bonding angle among adjacent bonds indicate that some of the molecules' parts become twisted. Simple bonds allow rotation around their covalent axis and create rotary isomeric forms. This possibility does not exist in double bonds.

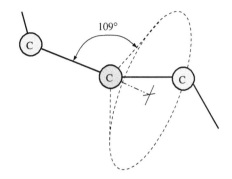

Figure 10.2 The carbon-carbon link in a curved chain as affected by spatial rotation.

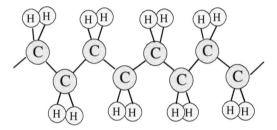

Figure 10.3 Spatial scheme of a piece of PE chain.

Second, if different functional groups are included in varied positions, there will be new spatial possibilities. In figure 10.4 two possible configurations for polypropylene (PP) can be seen.

Molecules are a construction of bonded rigid spheres with set angles and distances. Physicochemical constants can be obtained with enough approximation using this model. But oscillations gain amplitude around the equilibrium point, and the probability that various rotary isomers exist depends on the temperature. Hence these effects should usually be taken into account. Oscillations in bond angles are of the order of $\Delta\theta \approx 1°$–$10°$ at room temperature, which is small enough for short chains to make them insignificant. In short chains the most pronounced effect is their flexibility; this enables the bonds to be represented by the chain in figure 10.3. When chains are long, the rotational movements along them are summed up and clearly they become much more complex. The degrees of freedom deter-

Figure 10.4 Two possible spatial configurations for polypropylene.

mine the optical properties like the bands of absorption frequency and of radiation emission. To design a material with a predetermined behavior, many variations can be achieved; we work from the structure of the unit as we do with the macromolecule's architecture.

Modifications in the unit of a homopolymer generate an enormous range of possibilities in obtaining properties. The substitution of hydrogen atoms by other atoms or functional groups in the monomer's molecule creates drastic changes; the new functional group will contribute to the properties of the polymer as a whole.

Continuing with ethylene, figure 10.5 shows polymers obtained by substituting atoms of H in the original unit. If one hydrogen is substituted by a methyl group (CH_3), the PE transforms into polypropylene (PP). Although the degree of polymerization and the distribution of the molecular weight are approximately the same, the morphology of the macromolecule and its physical macroscopic properties will change. If the substitution is done with a benzene group (C_6H_6), the result will be polystyrene (PS). If it is substituted by chlorine, then we get polyvinyl chloride (PVC). These changes are so important that they stand out; for instance, PE is translucent, flexible, and crystalline, and PS is transparent, rigid, and amorphous. This does not mean that their synthesis was the same. Industrial methods of synthesis also depend on economic concerns, and their study is related to materials engineering.

A second possibility of structural modification is opened when contemplating the regularity and symmetry with which the new units are set in the polymeric chain (figure 10.6). This arrangement is defined by the synthesis method of the polymer, which also strongly affects the material's properties.

The head-tail configuration (figure 10.6[a]) is the most common because the repulsion energy is lower between the more distant radicals. The results of different configurations are called isomers, which are normally classified as *stereoisomers*, and isomers of the geometrical type (*cis* and *trans*).

Stereoisomerism

Stereoisomerism indicates configurations where the translational order inside the chain is retained but spatial order changes. For the disposition head-head, in which all groups are on the same side, we have a special case called isotactic isomerism (figure 10.7).

In figure 10.8, where groups alternate between one side and the other, the chain is translationally analogous to the previous type and is said to be syndiotactic.

When groups (say, methyl, in the example) have positions completely at random, they are said to be in an atactic configuration (figure 10.9).

Stereoisomerism is a major property of polymers, and it is looked for when the material is synthesized. For instance, PP is useful commercially only when it has an isotactic configuration.

A polymer material normally has a mixture of configurations that depend on the synthesis method. But stereospecific catalysts work on the polymerization mechanism.

The stereoisomerism is highly significant in relation to the resulting crystallinity, and in the dimensional stability versus temperature.

Geometrical isomerism

In instances where there is a double bond between the chain's carbon atoms, other configurations strongly modify the properties. One classic example is the polymer of isoprene,

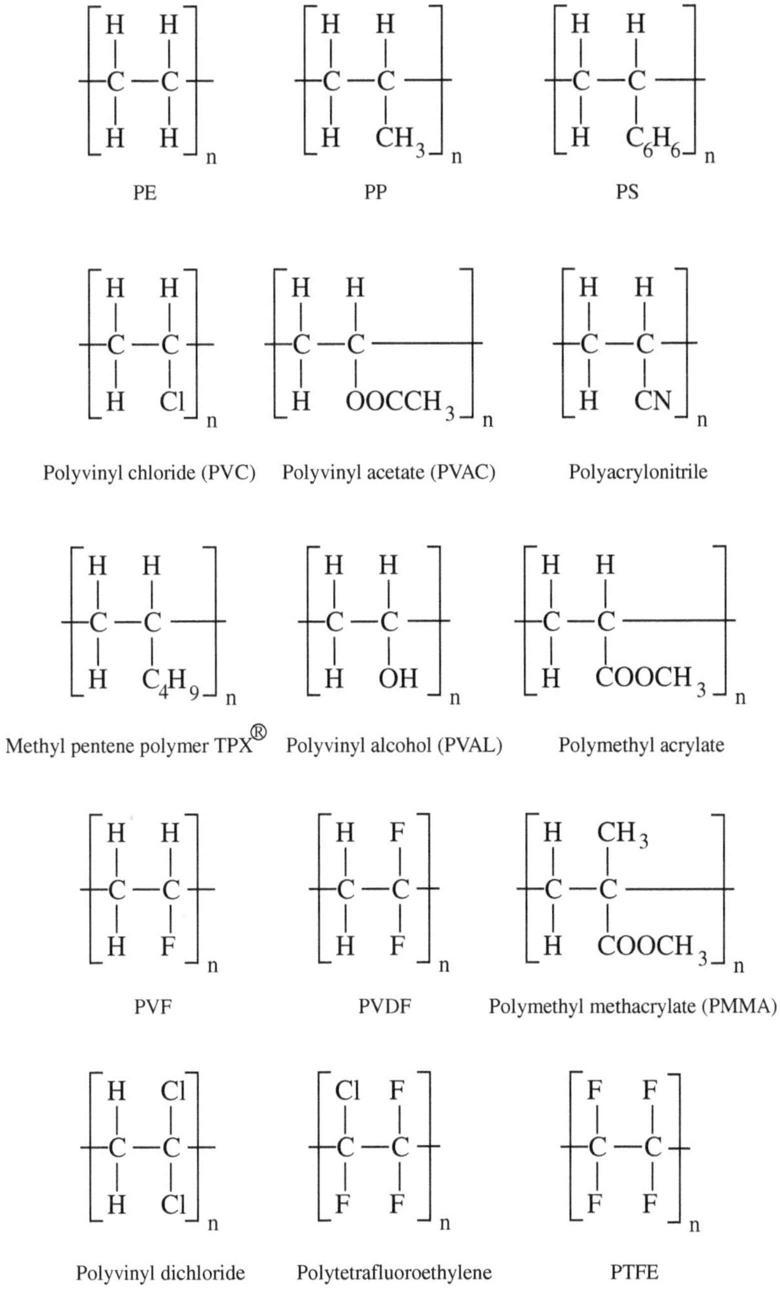

Figure 10.5 Polymers based on the schema of PE.

which has two configurations (figure 10.10) depending on whether the methyl groups and the hydrogen atoms are on the same or different sides of the chain. The first structure with both radicals on the same side is called *cis*-isoprene, commonly known as *natural rubber*. The second case, with radicals at each side, is called *trans*-isoprene, whose common name

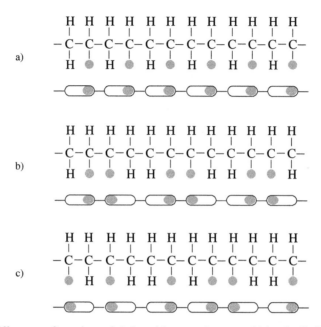

a)

b)

c)

Figure 10.6 Different configurations of chains with a generic group: (a) head-tail, (b) head-head, (c) random.

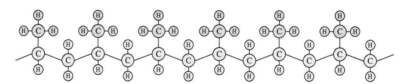

Figure 10.7 Stereoisomerism: isotactic polypropylene (PP).

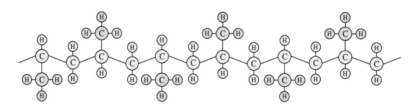

Figure 10.8 Stereoisomerism: syndiotactic polypropylene (PP).

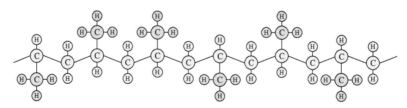

Figure 10.9 Stereoisomerism: atactic polypropylene (PP).

is *gutta-percha*; its properties are quite different from those of the former. It is not possible to change from *cis* to *trans* by simply rotating the chain because the double bond is rigid and does not allow it.

a)   b)

Figure 10.10 Geometrical isomers: (a) *cis*-isoprene (natural rubber) and (b) *trans*-isoprene (gutta-percha).

These have been examples of homopolymers, which share the property of having only carbon atoms in the main chain. We grouped them as carbopolymers. In this classification we also include other polymers that have oxygen in the chain, known as carboxypolymers (figure 10.11). We have added in the classification other polymers that include nitrogen as carboazopolymers (figure 10.12).

Polymers, as shown in the previous examples, can combine some nonexclusive structural elements on the same chain; for example, they can be linear and isotactic, or linear and have oxygen in the main chain. For that reason, the molecular structure must be specified with all details.

In instances where benzene groups appear in the main chain, some properties are strongly modified. These modifications take place mainly because the bond endows the skeleton with more rigidity when a material is attacked by chemicals; it also makes it tougher and fortifies its strength against impact.

Finally, there are macromolecular materials with no carbon atoms in the main chain. Among them are siloxypolymers, also called polysilozalanes or silicones. Their skeleton is formed of Si atoms that are bonded by oxygen bridges, and the free valences of Si are saturated by organic radicals. Polysilozalanes are normally synthesized from chlorosilanes. Depending on the nature of these chlorosilanes, other macromolecular architectures can also be obtained, like liquids (silicone oils), resins, or elastomers. All these materials have high thermal stability and are hydrophobic.

Heteropolymers or Copolymers

An alternative way to change the starting unit of the chain, as explained for homopolymers, consists of changing the chemical composition. This is accomplished by mixing different structural groups in a process known as copolymerization.

Here the starting point is from at least two different monomers—for example, styrene and acrylonitrile, which react to create the copolymer styrene-acrylonitrile (SAN). Again, many functional groups are possible if one chooses different starting monomers. Since two or more components can be used, the variations are many indeed. For example, the combination styrene-butadiene, with two components, can be modified by using acrylonitrile-butadiene-styrene (ABS) to achieve a three-component mixture.

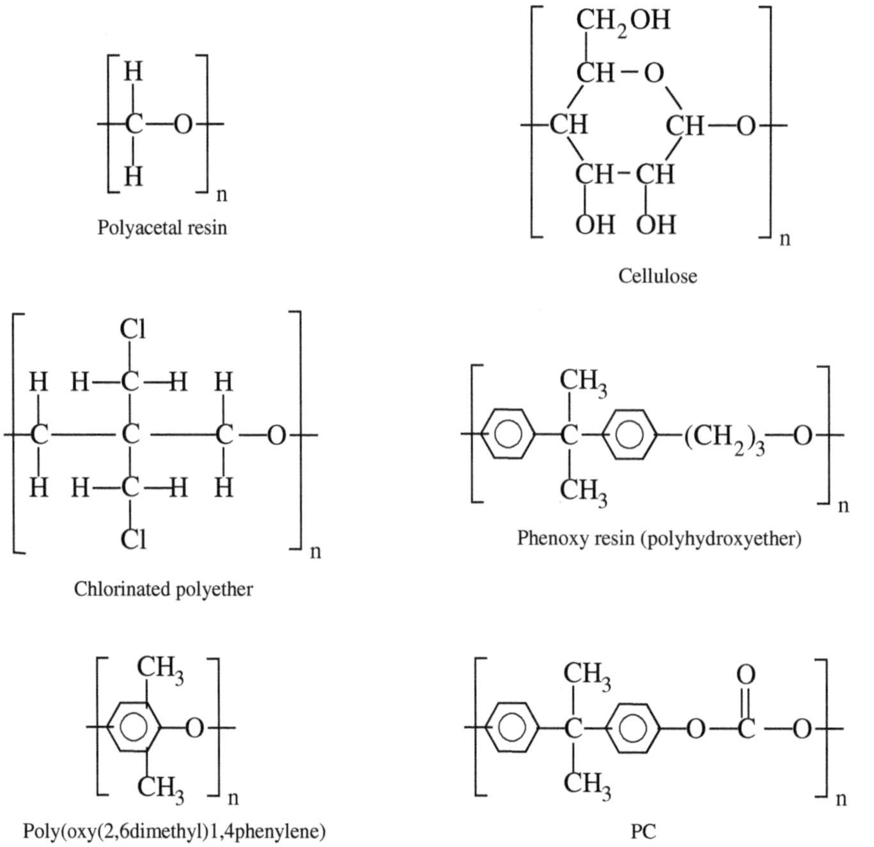

Polyacetal resin

Cellulose

Chlorinated polyether

Phenoxy resin (polyhydroxyether)

Poly(oxy(2,6dimethyl)1,4phenylene)

PC

Figure 10.11 Some polymers include oxygen (carboxypolymers) in their main chain, along with other groups.

If we change the proportions in the reaction, a nearly limitless number of combinations and properties, known by the coined phrase *plastics of design*, can be created with these mixtures (figure 10.13). In figure 10.13 the formulas miss an important bit of information: The same combination can be obtained with a different order inside the same chain (figure 10.14). Each species distributes itself in the chain according to the amount of reagents present rather than its polymerization. A copolymer could be obtained by prompting one component to react and then the other, or by letting both components coexist in the reactor and react freely.

10.2 MACROMOLECULAR ARCHITECTURE

Current techniques for synthesizing polymers allow a high degree of control over the structure of the molecular chains. We showed in the initial classification three shapes for these macromolecules: linear, ramified, and reticulated (figure 10.15).

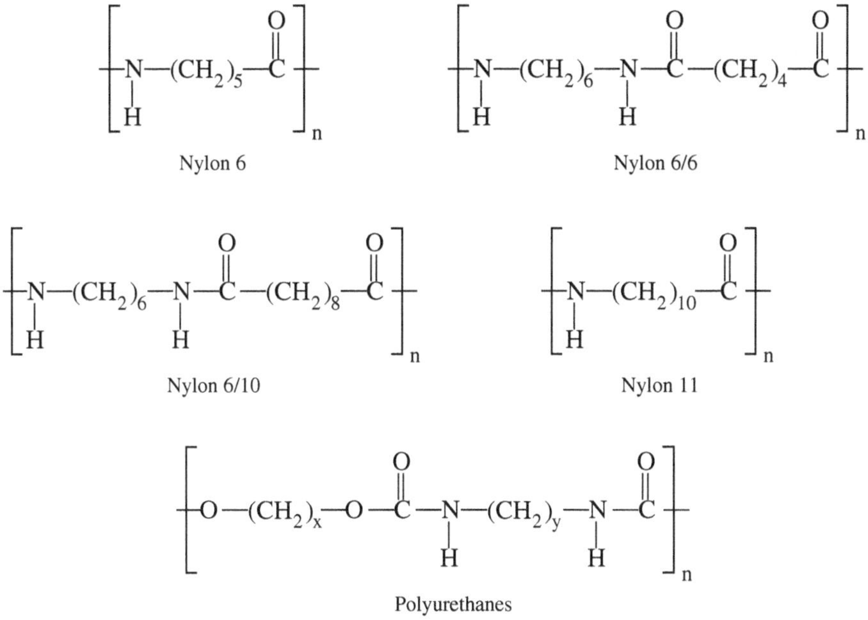

Nylon 6

Nylon 6/6

Nylon 6/10

Nylon 11

Polyurethanes

Figure 10.12 Polymers that include nitrogen in the main chain (carboazopolymers).

Order in Chains

First, there is no a priori reason to take for granted that the chains must be straight (apart from the zigzag of the macromolecule spine required by the angle of the carbon-carbon bond). A chain with a simple bond can lie in a plane, or can rotate and bend in three dimensions in an ordered way,[2] or can be disordered as shown in figure 10.16. In the example of PE, the bonds among carbon molecules must form an angle of about 109°, but there are several possibilities of ordering. In space, the first nearest neighbor of a carbon atom, placed at any point on the chain, can appear at any point on a circle, as indicated in figure 10.2, that preserves the distance and bond angle. This allows the chain to bend and wind to create a three-dimensional coil. One of the first consequences is that the net distance among the chain's extreme atoms is normally a lot smaller than the total length of the chain. As we will see, statistical measurements of the relation between net length and apparent diameter yield a small number.

Many important properties, such as the packing, which defines the density or the elasticity in rubber, are due to this type of winding of the molecular chain. The capacity of chains to rotate also affects properties like thermal vibrations or the response to stresses. This ability to rotate is defined by the monomer's chemical structure, and it is obstructed by double bonds (C=C) or by the substitution of H atoms by other atoms or chemical groups (e.g., one benzene ring, as in PS). The closeness of other chains with bulky groups can also obstruct the rotation capacity in a polymeric chain.

---

[2]The most important example in nature is the double helix structure of DNA.

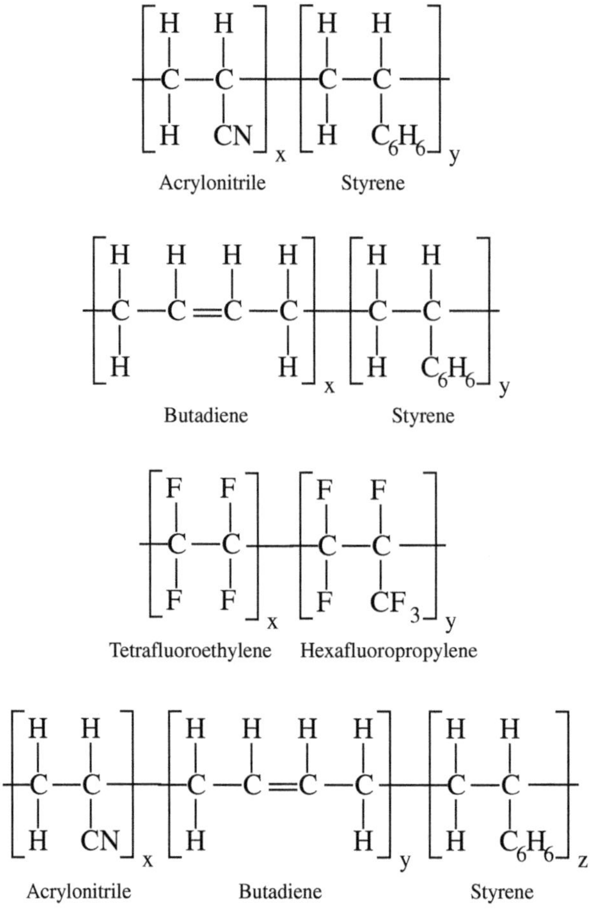

Acrylonitrile   Styrene

Butadiene   Styrene

Tetrafluoroethylene   Hexafluoropropylene

Acrylonitrile   Butadiene   Styrene

Figure 10.13 Copolymers of two and three components. In each group the x, y, z subindices represent the fractional composition of each component.

Branching

In a similar way to copolymers, where the situation is obvious, the macromolecules of homopolymers have branches. The branching is controllable by using catalysts and changing the reactor's pressure during a polymer's synthesis. Branching, like other properties, reduces the density, which becomes dependent on the number of branches per unit of length; so compact packing becomes more difficult. For example, creation of one important branching every 500 units and smaller ones every 50 units during the synthesis of polyethylene yields low-density polyethylene (LDPE).

Cross-Linked Structures

When adjacent lateral chains are joined by bonds, important forces act among the chains, which can be of the van der Waals type, hydrogen bridge bonds, or covalent bonds, and an interlaced or networked structure is formed, either two or three dimensional. Many elastic materials like rubber become interlaced during vulcanization. Bridges between

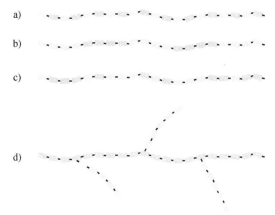

Figure 10.14 Order of polymers in chains, where each type of shading represents a type of copolymer: (a) alternating, (b) in groups, (c) grouped at random, and (d) branched.

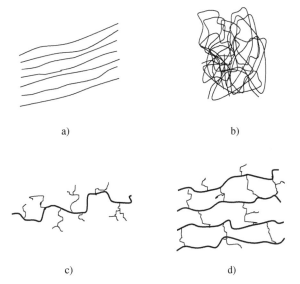

Figure 10.15 Macromolecular architecture: (a) linear (parallel), (b) linear (entangled), (c) branched, and (d) networked.

chains can be generated during synthesis or by irreversible chemical reactions, normally at high temperatures. When the monomer has three active covalent bonds (three-functional monomers) the structure is three dimensional and hence differentiated from linear structures formed by bifunctional groups. Epoxy and phenol-formaldehyde resins are examples of such polymers.

Chain Length

One aspect of macromolecular substances is that they are a mixture of similar macro-molecules with different degrees of polymerization. The differences are so small that they

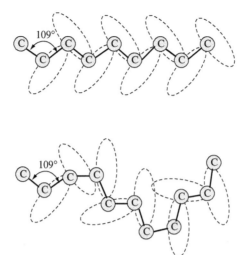

Figure 10.16 Scheme of a straight chain (a) and of a chain twisted due to spatial rotation (b). See also figure 10.2.

cannot be separated by normal methods used for diverse individual species. The distribution of the macromolecules' degree of polymerization is centered on a mean value $\langle n \rangle$, which implies that this distribution is true for the molecular mass M. Hence, these substances are called polydisperse.[3]

Despite this macromolecular distribution, high polymers behave as one unique substance. The consequences of the statistical distribution are much more significant than the differences in physical properties. Among other consequences derived from their being formed by disperse molecules, they lack a well-defined melting point. High polymers have a zone of melting or softening due to their being formed by molecules that differ in size and because they possess secondary forces that cause complex interactions in a polymeric material.

Many polymeric materials—in particular, those that have molecules of a linear type—have crystallized areas where the fibers have a parallel order, like fibrils, which coexist with disordered areas as in amorphous materials (figure 10.17). Cellulose, polypropylene, and so on are examples.

10.3 CRYSTALLINE, SEMICRYSTALLINE, AND AMORPHOUS POLYMERIC MATERIALS

According to what we learned in earlier chapters, we know that the crystalline states in metals or ceramics correspond to an ordered disposition of atoms or ions. With polymeric materials the crystalline phase may also exist, but this order concerns the disposition of atoms that belong to a macromolecule; thus it is a more complicated situation. Figure 10.18 shows the possibilities that chains which form the macromolecule have to get in order. Crystalline orderings can be observed in chains that try to place themselves in a parallel way. Orderings range from a high degree of crystallization to an amorphous solid or viscous liquid, depending on how the chains are arranged.

---

[3]However, one must appreciate that natural substances, the proteins, are formed by only one type of macromolecule.

Figure 10.17 Forming fibrils by stretching chains.

a)   b)   c)

Figure 10.18 Scheme of (a) crystalline, (b) semicrystalline, and (c) amorphous ordering.

Many polymeric materials, especially those with linear molecules, display crystalline areas where fibers are arranged in a parallel fashion with ordered fibrils, and disordered areas, which is like the arrangement of amorphous materials. Examples are cellulose, PP, and so on.

In linear chains crystalline order is easier to achieve, but bulky groups replacing H atoms in carbopolymers or the presence of branching prevent a high degree of crystallization. Consequently, the degree of crystallization depends more on the chemical composition than on the chain configuration. Crystallization is not favored in a chain formed of chemically complex units, and it is difficult to avoid in polymers that have simple chemical structures.

Achieving a degree of crystallization in a controllable fashion by the manufacturing process is something that differentiates polymeric materials from metals or ceramics. In metals, the usual goal is to obtain a completely crystalline phase; in ceramics one desires either completely crystalline or completely amorphous. A polymeric material's situation is more similar to a two-phase metallic alloy. This different ordering has a great deal of influence on the properties.

In figure 10.19 order can be found in a unit cell of the PE variety. The cell has orthorhombic ordering in one region of the chain. It is delimited by the dotted line $(0.255 \times 0.494 \times 0.741 \ nm^3)$. If this order were possible over the whole sample, we would have a perfect crystal. But not all chains have the same length. They are equally aligned, but they suffer torsion and foldings, which create amorphous zones. This makes the crystalline regions spatially scattered, coexisting with other amorphous regions and originating a partial crystallization. The corresponding material is semicrystalline. This affects, first of all, the density. If the crystal were perfect, the degree of packing would be higher and hence the density would be too. When the material becomes more amorphous the density lessens. Based on density measurements, a material whose density is ρ can be assigned a mass degree of crystallization (%); this refers to the density of the same polymer that is completely crystalline ρ_c and completely amorphous ρ_a:

$$c\ (\%) = \frac{\rho_c(\rho - \rho_a)}{\rho(\rho_c - \rho_a)} \cdot 100.$$

Controlling the degree of crystallization in a polymeric material is accomplished by acting on the solidification (section 9.2) through a saturated solution, or the polymer's viscosity, or the chain's configuration.

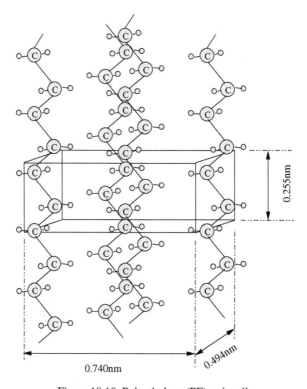

Figure 10.19 Polyethylene (PE) unit cell.

Different models have been offered as explanations of why chains are disposed to form semicrystals. Among these models are the following:

• A model of crystallites (micelles) with fringes.

• A model of a folded chain.

The first model, which was accepted and used for many years, consists of a polymer where a crystallite is formed by a group of chains partially aligned and embedded in an amorphous matrix, as shown in figure 10.18. Regions of the same macromolecule belong to a crystallite or to an amorphous zone. But studies of crystalline growth from dilute solutions have revealed that the crystals usually consist of thin plates of 10–20 nm thickness and 10 mm length. These crystals, presented in figure 10.20, can be readily observed in a scanning tunneling microscope (STM). This type of ordering is known as a folded chain; a chain can occupy a thickness of 10–20 nm before it folds on itself, whereupon the action repeats until the entire length is folded. Because the length is a lot bigger than its thickness, this gives the crystal's direction a superior size. One plate can contain several molecules, but the crystallite's length is much larger in one direction.

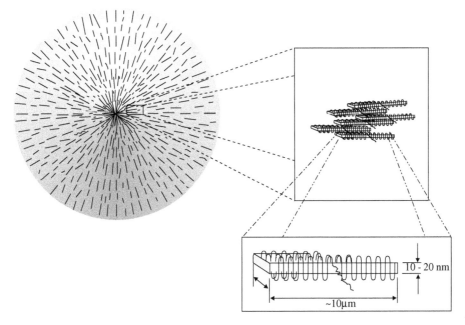

Figure 10.20 Formation of spherulites through lamellae.

Several investigators (e.g., R. Boyd and J. Corburn) have revealed that these plates or lamellae in polymeric materials crystallize from a liquid phase and group in small *spherulites* (figure 10.20) equivalent to a grain of a ceramic or a polycrystalline metal. Lamellae connect the spherulites through amorphous regions. Examples of such order that is crystallized from a liquid are PE, PP, PVC, and polytetrafluoroethylene (PTFE), among others. Observing the samples with a polarized light microscope, we can see in each spherulite a maltese cross, typical of a birefringent crystal in the direction of the optical axis.

10.4 DISTRIBUTION OF MOLECULAR WEIGHTS AND SIZES

Molecular Weight

The molecular chains that compose high polymers normally have different lengths. Even if the molecular composition is maintained, uncontrolled factors in polymerization affect the length of the new molecular chains. Even if we keep the reactor's parameters constant, we arrive at a continuous distribution of chains with different lengths. This prevents the macromolecules from being uniform—described either by a unique value of molecular weight or by how many units are combined in the chain. The result? Both change according to the chain length. The molecular weight and the number of units in a sample are typified by their mean value and by their dispersion.

Many macroscopic properties depend on this distribution. To obtain the curves experimentally, it is convenient to start from a dilute solution, in which the macromolecules are more separated and their interactions are not so important. Using appropriate methods, researchers group the polymer's fractions according to molecular size, they measure the number or weight, and then they represent these data along a distribution curve. The curves depend on the measurement method; that is, they differ if the method responds essentially to the molecular weight or to how many particles there are by volume unit. For example, experimental methods that chemically determine the final group or the osmotic pressure are proportional to how many particles are present; methods that respond to viscoelastic properties are proportional to the molecular weight. Some methods and modifications allow one to obtain other averages (M_m, M_n, M_z) by electron microscopy or X-ray diffraction. Figure 10.21 presents distributions with different dispersions.

With the values obtained experimentally the following magnitudes are defined:

- Numerical average $M_n = \sum n_i M_i / \sum n_i$

- Mass average $M_m = \sum m_i M_i / \sum m_i$

- Zeta average $M_z = \sum n_i M_i^3 / \sum n_i M_i^2$

If a substance had equal-sized molecules, the three values would obviously be the same. Since it is not like this and the result depends on the kind of measurement, the values are slightly different. The mean degree of polymerization $\langle n \rangle$ is obtained by dividing the averages by the monomer's molecular weight. The bigger the dispersion of molecular sizes, the more the mean values with each method differ. This difference allows one to predict polydispersion in a sample.

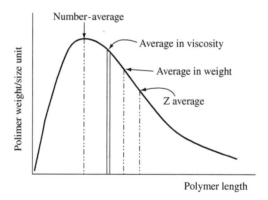

Figure 10.21 Sketch of a molecular weight distribution curve.

Methods for Obtaining Molecular Distribution Curves

Determination methods are normally classified according to their origins:

1. Methods that use *colligative properties*. The numerical average M_n is obtained within the limits inside parentheses.

Ebullioscopy ($< 4 \times 10^4$)

Cryoscopy ($< 5 \times 10^4$)

Osmometry (10^4–10^6)

2. Methods based on *optical properties* or direct observation. In these methods the mass average M_m is obtained, except for the electron microscope, which permits the estimate of M_n.

Light scattering ($> 10^5$)

Diffraction (10^4–10^7)

Double refraction (10^4–10^7)

Electron microscopy ($> 10^6$)

3. Analytic methods that use the *determination of the final group*. In linear polymers each molecule has two ends. By determining how many terminals are in a sample, the number of macromolecules can be obtained as well as the molecular weight average M_n. Terminal functional groups are identified through analytical chemistry techniques. One large complication is due to branching, which modifies the number of terminal groups in a macromolecule and hence creates errors. For this reason, this method, when combined with others, gives an idea of how much branching is present. Determinations of M_n are obtained up to 2×10^4.

4. *Mechanical* methods enable us to obtain the mass average M_m, except for the method of sedimentation equilibrium, which has an average of M_z.

Viscosimetry (without limit)

Sedimentation rate (10^4–10^7)

Sedimentation equilibrium ($> 10^5$)

With modifications, the sedimentation equilibrium method can reveal other averages and enlarge the interval of molecular weights (modifications of Archibald, which is handy to have M_z and M_m, and Trautman, which is useful in obtaining M_m).

Statistical Length and Radius of Gyration

The properties of a polymer normally depend on the molecular mass as well as on the macromolecular shape. Thus, it is not enough to know the distribution curve of molecular masses; it is also necessary to know the distribution of macromolecules among the shapes that a polymeric material can adopt. The possibility of taking on different shapes and sizes usually causes differences in some properties in a sample. Determining the curve of molecular distribution is easy with several methods. Determining the molecules' shapes is more difficult.

When the degree of polymerization is increased, the agglutinative power grows and the tendency to form fibril groups with long molecules having minimal branching generates a strong inclination for crystalline regions to appear. When lateral chains appear, crystallization gets into difficulty. It is clear that for a certain molecular weight and number of monomers, the size of a highly branched molecule will be smaller than that of less branched or linear ones. A stretched molecule formed by n monomers of length a, linked at a valence angle α kept in a plane, would have a length expressed by

$$l = na \sin\left(\frac{\alpha}{2}\right).$$

This is the ideal limit.

Different statistical models enable one to estimate more realistic situations in space. For example, a spatial mathematical model would permit one to calculate a more realistic shape in three dimensions and project it on a plane as in figure 10.22. One classical model, called the self-avoiding random walk, is built by estimating the path at random, according to the number of steps, and by avoiding junctions that constitute unrealistic situations. Here molecules form balls and have lateral mean sizes r_0 comparable to the longitudinal one. Experimental measurements confirm these comparable sizes (length $\approx$ 2–6 times the radius of gyration r_0). The most complete models of statistical mechanics predict that the medium transverse size is approximately half the longitudinal one; that is, one macromolecule will have a high probability that it is twice as long as its width.

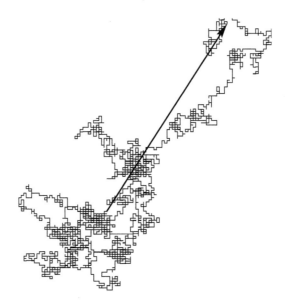

Figure 10.22 Numerical simulation of an isolated polymeric chain by a self-avoiding random walk (courtesy of D. L. Valladares).

The ball can have a mean diameter d (expression [10.1] below) based on its longitudinal size and related to the monomer's statistical size a_0 ($a_0 > a$). A semiempirical relation is normally used for molecules in solution, whose coefficients are adjusted by values obtained through measurements related to size: light scattering, sedimentation, diffusion of the polymer's dissolution, and, partially, viscosity:

$$d = a_0 n^{(1+e)/2}. \tag{10.1}$$

In this expression, the exponent e depends on the solvent, and its value fluctuates between 0 and 0.333, being bigger when the solvent has more power. When using a poor solvent the value is close to zero (a precipitating solution) and the expression can be simplified independently of the solvent used:

$$d = a_0 n^{1/2} \approx M^{1/2}.$$

The value of d increases monotonically with n (or with M), which means that when raising the molecular weight, the macromolecule does more winding. The degree of winding can be measured with the *Kuhn coefficient*, which relates length to diameter: l/d. This coefficient increases with the square root of the degree of polymerization or with the molecular weight:

$$l/d \approx \frac{n}{n^{1/2}} \approx n^{1/2} \approx M^{1/2}.$$

The whole expression (equation [10.1]) represents the efficiency of a solvent for a polymer and verifies whether the macromolecule is stretching in solution, as already checked experimentally. Stretching also happens when the temperature rises—a key part of the analysis of a manufacturing processes.

If the molecules are branched, for a particular molecular weight they will have less extension than ones that are not branched; thus, there will be a difference in the mean diameter. It would be possible to infer from this statement a method to determine the number of branches, but unfortunately a measure of the mean diameter allows only a qualitative knowledge.

If the molecules are reticulated or have strong cross-links, the chains are not differentiated, and hence the group can be judged as only one macromolecule whose molecular weight could, in principle, become the sample's entire mass.

10.5 POLYMERIZATION PROCEDURES

Most monomers that constitute the raw materials for the synthesis of macromolecules in high polymers are derived from petroleum, coal, and natural gas. These starting substances are obtained physicochemically by cracking, distillation, extraction, and so on. They are the basis of a huge variety of synthetic plastic materials. Products that play major roles in manufacturing monomers are tar and coke, which have the range of aromatic products, and acetylene, from which unsaturated aliphatic products are derived.

The monomers described in section 10.1 are unsaturated substances whose molecular weight is low, but after opening their double bonds or cyclical structures they become bonded in macromolecular chains and form known materials. Many final properties depend on the starting monomer. Some are controllable during polymerization or transformation of a polymer. For example, including aromatic groups in a macromolecule's main chain enhances its rigidity.

In any process of polymerization there are three stages: the beginning of the reaction, the growth or propagation of the reaction, and the breaking or end of the reaction. The reaction starts with the activation of the double bond, like the preceding process for creating a chain. The reaction to breaking interrupts the growth. Compounds with nitrogen, phenol, ferric salts, and so forth are inhibiting substances. Even when created by the same substance, inhibition can be either complete or partial.

In polymer engineering several procedures achieve polymerization (in blocks, in beads, in suspension, in dissolution, in precipitation, in emulsion, in dispersion, etc.) and the details are analyzed by materials engineers. Monomers have different tendencies to polymerize (or to co-polymerize), and these depend on the larger or smaller polarity of the double bond and other factors. Although any method of polymerization would be usable for a monomer, normally one procedure is the most efficient. Let us consider the reactions that summarize industrial methods: polymerization itself, condensation polymerization, and adduction polymerization.

Chain Polymerization

In polymerization itself (also called addition or chain polymerization), the chaining of molecules happens without separating the simple molecules. Therefore, the polymer's final composition is the same as that of the monomer that originated it. Polymerization can be formulated as

$$n(CH_2 = \underset{\underset{R}{|}}{C}H) \longrightarrow -[CH_2 - \underset{\underset{R}{|}}{C}H-]_n$$

According to the two types of end structures that can be formed by heterolytic and homolytic breaking of double bonds, polymerization can take place according to either an ionic or radical mechanism.

The double bond C=C is formed by a σ bond and a weaker π bond. The chemical characteristics come from the π bond. π electrons can move more easily than can σ electrons because their charge is linked less strongly to the carbon atoms by the screening of the σ electrons. Through an external action, the π electron pair can be moved until it belongs to only one carbon atom. This generates two-ended polar structures created by a heterolytic split (figure 10.23).

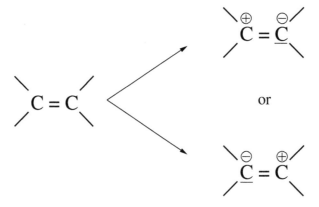

Figure 10.23 Polar structures derived from the heterolytic splitting of the π-electron pair.

The other possibility consists of activating the double bond by homolytically splitting the π electron pair. Now an end structure with two radicals is formed (figure 10.24).

This anomalous distribution of electrons is decisive for this type of compound to produce polymerization reactions.

The activation of the double bond can start by the action of light, heat, ultrasound, radical formation, or catalyst agents. The radicals being formed are substances that, by the action of heat, decompose easily into free radicals, like diacetyl or dibenzoyl peroxide,

Figure 10.24 Activation of a double bond by homolytically splitting the π-electron pair.

perbenzoic acid, compounds of nitrogen, and so forth. The catalysts (acids or bases) are substances that operate only at the beginning of the reaction; they do not appear in the final products.

In a reaction started by radicals (acids or bases) polymerization advances according to the ionic mechanism; whereas in one activated by light, heat, ultrasound, or radical formation, polymerization develops via the radical mechanism. The propagation of the reaction is fast (from 0.01 to 0.001 s for every 1000 units of monomer). The reaction can end in different ways; for example, when both active extremes react and generate an inactive molecule, or when an active extreme reacts with the initiator or other chemical substance with a simple active bond. The relative speeds of initiation, propagation, and ending determine the molecular weight, and their controllability opens the way to deriving a certain molecular weight. Since the reaction's ending is random, chains will have different lengths and thus cause a distribution of molecular weights.

Chain polymerization is the usual mechanism in polyolefins (PE, PP, PVC, PS) and in copolymers generated by butadiene, vinyl acetate, styrene, and acrylonitrile.

Condensation Polymerization

In polycondensation, which is also called polymerization by stages, polymers are formed from more than one kind of monomer; normally a low-molecular-weight by-product, which should be eliminated, is produced. Therefore, the resulting polymer has a final composition different from that of the monomers.

If the initial monomers are bifunctional, the polymer will be a linear condensate. If we have polyfunctional monomers (tri, tetra, etc.) three-dimensional reticulated polymers are formed. By combining bifunctional with polyfunctional monomers, branched polycondensates are formed. These can give rise to reticulates if the proportion of the polyfunctional monomer is sufficient.

The linear and branched polycondensates are thermoplastics, whereas reticulated ones are thermosets.

One classic example is the creation of a polyester from a saturated dicarboxylic acid such as succinic acid and a dialcohol such as ethylene glycol. The amount of water obtained as a by-product is normally a measure of the reaction's magnitude. The reaction can advance until the lowest possible monomer is consumed. Another classic example is the acquisition of alkyd resins, polyesters formed by condensing a bifunctional acid like phthalic or maleic acid with a triol such as glycerine.

Three-dimensional polycondensates can be obtained by the branching of a linear polycondensate. For this to happen the chains must retain some double bonds.

Among the best-known polycondensates are the families of phenoplastics and aminoplastics (called formaldehyde or formolic acids), polyesters, polyamides, tioplasts, polysulfurs, and silicones. From reactions of formaldehyde with phenol, proteins, urea, aniline, and melamine we derive plastics that are hardened by the effect of the heat; that is, thermosets.

Adduction Polymerization

Adduction is the third mechanism by which polymers are created. It resembles the first two mechanisms. As with the first mechanism, the molecules of a polymer have the same eventual composition as the starting monomers (i.e., there is no separation of low-molecular-weight substances). Yet, given the kinetics, polyadduction is more similar to condensation. As in polymerization by condensation, the starting monomers are generally bifunctional or polyfunctional. But here in each stage of the reaction hydrogen atoms migrate from one functional group of a monomer to another. This adduction activity is different from condensation. Valences that remain free in this transfer bond the molecules.

Unlike the first two mechanisms, the groups react among themselves and cannot be in the same molecule. Polyadduction in chains with only one type of monomer is not known. Reactions usually take place at room temperature and do not require catalysts. The most common examples are epoxy resins and polyurethanes. During the hardening reactions of polyester resins, polyfunctional amines are needed for branched products, and organic acids are sought in less branched products (as in the varnish industry).

Polyurethanes are derived by adding polyfunctional isocyanates to polyfunctional alcohols (polyesters and hydroxylated polyethers). Expanded polyurethane is commonly used as an insulator and in the packing business. The CO_2 released is captured during reaction in the presence of water; the polyurethane formed during the simultaneous reactions of decomposition and addition acts as a sponge. The captured gas isolated in small bubbles cannot transport heat by convection; thus the material is a good thermal insulator.

10.6 PROPERTIES OF POLYMERIC MATERIALS

Polymeric materials with industrial applications were initially classified as thermosets, thermoplastics, and elastomers, alluding clearly to their thermomechanical properties, initially the most relevant. The strong dependence on temperature of a polymeric material's properties makes it essential to highlight their thermal behavior.

From the microscopic viewpoint the properties of *thermoset* materials such as phenolic resins (the first plastic to exhibit high mechanical and thermal resistance) are better understood by analyzing their structure of entangled chains and their degree of crystallization. In the entanglement the chemical bonds among chains normally happen when a polymeric material is heated during the final stage of its manufacture. This creates a three-dimensional web that obstructs any new heating process, and so a new molding cannot be done. This behavior explains the name "thermosets."

With *thermoplastics* another effect is important: to soften enough when heated to permit their molding by the different means studied by materials engineers. The term "plastic" means able to deform in a permanent way and thus able to be shaped under pressure and appropriate temperature. Here strong bonds are not formed between polymeric chains, although they can interact among themselves. Upon cooling the material stays amorphous as it takes on the aspects of a glass.

The entanglement that gives stiffness to thermosets (albeit with definite differences) is shared by synthetic and natural rubbers. But this time it is induced to obtain more elasticity. This disposition to entangle is exploited to enhance the material's resistance to temperature and to enable it to bear repetitive elastic deformations without being deformed irreversibly. These materials are known as *elastomers* or rubbers.

Many properties of polymeric materials depend on the molecular weight and the degree of crystallization. Figure 10.25 shows the separate regions typical of oils, waxes, and polymeric materials in the solid state.

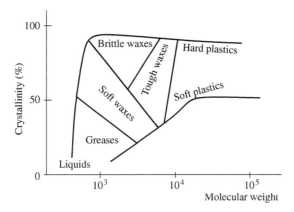

Figure 10.25 Sketch of a phase diagram for oligomers and polymers of ethylene.

Thermal Properties

Thermal properties are clearly affected by the structure of polymeric materials: polymeric bonds are much more dispersed than are bonds in other solids. This makes the thermal conductivity small, because there is little transmission of vibrations, and the specific heat is huge, because the molecular vibrations are independent, in contrast to other materials. They are, therefore, classified as thermal insulators. The more densely packed or *crystalline* the material, the higher is the conductivity and the lower the specific heat becomes. In the vibrations of the material, both whole molecules and pieces of molecules have mobility—an important difference compared to other thermal insulators. The two kinds of mobility determine macro-Brownian and micro-Brownian movement.

Melting Point (T_m) and Glassy Transition (T_g)

In polymeric materials where the chain structure is linked to amorphous materials, whose effect is added to the statistical distribution of molecular weights, the phase transitions are not narrow and sharp, but are characterized more by temperature intervals. As with ceramics, a major difference arises between crystalline and amorphous materials—in particular, with reference to the phase transitions. Thus, some highly crystalline polymeric materials melt at a defined temperature T_m—in particular, ones with low molecular weights. Most of them change to a phase similar to a viscoelastic amorphous solid.

Other polymeric materials, with a strong, branched structure of covalent bonds, do not melt directly but decompose in an irreversible way. The cross-links between chains that keep a structure stiff cannot reform after being broken. Decomposition normally takes place at such low temperatures that these materials have to be used at temperatures lower than the melting point of the crystalline phase. In the decomposed phase they behave as a gel: a suspension of a solid in a viscous and amorphous liquid phase. This phenomenon is not seen in some faulty products of PVC, where, after hot molding, one sees a surface roughness that comes from the original material in the solid phase.

The qualitative thermal behavior of those polymeric materials that melt without decomposing is as follows. At low temperatures they are brittle and stiff—in a glassy phase. When the temperature rises they acquire elasticity—a viscoelastic phase, as depicted in figure 10.26—before turning into a liquid phase. This transition is often explained by assuming that the lateral chains start acquiring mobility first; this is called micro-Brownian movement. The transition from the liquid phase of frozen micro-Brownian movement to free micro-Brownian movement is called the *glassy transition*.

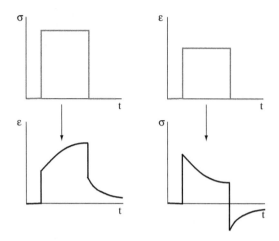

Figure 10.26 Sketch of the behavior of a viscoelastic phase under the exertion of constant stress and strain.

During a subsequent rise in temperature, the cohesive forces that keep the molecules bonded are overcome and movement increases (macro-Brownian) until the molecules move freely and the material softens.

Finally, the material reaches the liquid phase, where thermal agitation is intense enough to break the main valences and the macromolecule decomposes without proceeding to the gaseous phase. A convenient way to illustrate this is to analyze and compare the curves that represent volume in relation to temperature (figure 10.27).

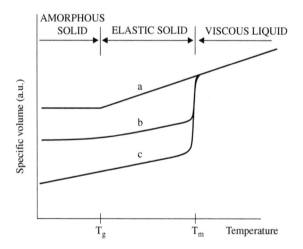

Figure 10.27 Thermal behavior of a polymeric material as a function of crystallization. T_g is the glassy transition temperature and T_m is the melting point temperature. (a) amorphous, (b) semicrystalline, and (c) crystalline material. T_g has no meaning in crystalline materials.

Amorphous solids exhibit a continuous gradation in the specific volume when heated until they reach the liquid phase. But when they cool, the curve has a change in slope when the solids reach a value of temperature T_g, the glassy transition temperature, which generally depends on the cooling rate (see section 9.2). If we continue cooling below T_g, the material behaves as an amorphous solid. Above T_g, when the temperature rises the material is first a soft solid and then an elastic solid and finally a viscous liquid. In amorphous polymeric materials, the clear presence of two scales of movement (micro- and macro-Brownian) makes the glassy transition independent of the cooling rate.

Semicrystalline polymeric materials, whose curves are intermediate, have the properties that characterize both the amorphous and crystalline phases, that is, a glassy transition and a melting temperature.

Specific Heat and Thermal Conductivity

Specific heat and thermal conductivity are strongly conditioned by the structure of the polymer. According to chapter 5, the high thermal conductivity of metals is related to free electrons. With polymeric materials like macromolecules, the bonds are much more widely dispersed than in other types of solids. They have low thermal conductivity (table 10.1) and with few free electrons there is low transmission efficiency of vibrations because of so much phonon dispersion. In the same way they have a large specific heat (table 10.1) due to their formidable capacity to produce vibrations independent from each other—above all, amorphous polymeric materials. As a consequence of this structure, when polymeric materials become more and more crystalline and approach closer packing, the thermal conductivity increases and the specific heat decreases.

Table 10.1 Thermal properties of common polymeric materials

Material	T_g (°C)	T_m (°C)	c_p (J/gr · K)	κ (J/m · s · K)
Polyolefins				
Polyethylene (PE)	-120	133	1.81	0.2
Polypropylene (PP)	-17	187	1.71	0.2
Polyisobutylene	-73	117	1.665	0.18
Vinyls				
Polyvinyl chloride	81	273	1.10	0.18
Polyvinyl fluoride	-18	230	1.35	0.168
Aromatics				
Polystyrene (PS)	100	243	1.21	0.16
Polycarbonate	148	230	1.135	0.15
Nitriles				
Polyacrylonitrile	105	318	1.24	0.20
Acrylates				
Polymethyl methacrylate	105	200	1.39	0.31
Polyamides				
Polyheximethylene (nylon 6/6)	43	226	1.53	0.23
Polyacrylamide	163	304	1.22	0.15

The ratio between the thermal conductivity κ and the product of the density ρ with the specific heat c_v is called the thermal diffusivity χ. This ratio indicates how fast a sample cools at room temperature. Although the density of polymeric materials is small, their thermal diffusivity is also small and it requires more time to cool a material than the sample can hold up under its own weight without deforming. This has major technological implications for molding by injection or extrusion.

Mechanical Properties

Density

The density of polymeric materials is distinguished by being low and by increasing with the degree of crystallization. The density of the commonest thermoplastics changes from values smaller than that of water to those of polyolefins, which are quite high because these polymers contain halogen groups. Other polymers have intermediate values, as shown in table 10.2.

The density of PE with a particular degree of crystallization can be estimated by means of a linear law that combines the density of the amorphous phase ($\rho_a = 0.85$ g $\cdot$ cm^{-3}) and the density of the crystalline phase ($\rho = 1$ g $\cdot$ cm^{-3}). The melting of PE causes a great increase in the volume, and consequently the density changes to values lower than those corresponding to the amorphous phase (e.g., 0.75 g $\cdot$ cm^{-3}). This makes it necessary to maintain high pressures during cooling after molding, in order to avoid superficial sinking and hollows in the bulk.

Table 10.2 Density of some common polymeric materials

Material	Structure	Crystallization	Density
PP			0.85–0.92
Commercial PE			
Low density	linear, branch	$\approx$50%	up to 0.92
Low density	linear, little branched	50%	0.92–0.94
High density	linear, little branched	90%	0.95
PTFE			2.1–2.3

Elasticity

The elastic and plastic behavior of polymeric materials is different in some respects from that of metallic solids or ceramics due to their sensitivity to temperature and to the wide variety of values and behaviors. To characterize this behavior, stress-strain tests analogous to those for other materials are used. In polymeric materials, however, additional parameters participate: deformation speed, temperature, and the chemical influence of the medium (solvents, water, oxygen), which modify the results. In addition, when materials are elastic other tests are more convenient.

It is important to highlight that, in addition to this peculiarity, the mechanical behavior varies greatly according to the type of polymeric material (figure 10.28).

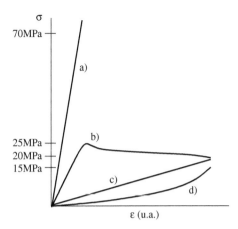

Figure 10.28 Stress-strain curves for (a) polymethyl methacrylate ($-40°$C), (b) polymethyl methacrylate ($120°$C), (c) vulcanized natural rubber, and (d) nonvulcanized natural rubber.

Thermoplastics behave like concentrated polymeric solutions with high viscosity. The main component of viscosity in thermoplastics, is the viscosity due to the axial creep of the polymeric chains. This viscosity, because the chains are long, is high even in normal conditions of temperature and pressure and can be modified by applying pressure or heat.

The viscosity in thermoplastics also depends on the rate of deformation, although the range of the rate of deformation is relatively low and this viscosity is almost constant. In this situation the thermoplastic is a Newtonian fluid. It is also a Troutonian one, characterized by extensional viscosity[4] proportional to the shear viscosity. The shear viscosity in this regime is proportional to the mean molecular mass to the power of $7/2$. When the rates of deformation are high, the thermoplastic viscosity is usually reduced, as with low-density polyethylene. Also the fluid is no longer Troutonian. Because this effect is reversible, if the rate of deformation is reduced again, the initial viscosity is restored.

In their initial stage elastomers are thermoplastics, but even at low temperatures they keep their amorphous or glassy phase. These are the gummy thermoplastics which can be stabilized by forming networks. For example, natural rubber comes from latex, a viscous fluid. But its stereoisomer, *trans*-polyisoprene, is gutta-percha, where the interference among CH_2 groups inhibits the flow. Latex can be stabilized by being heated with sulfur (a process called vulcanization), which forms a network in the material with short chains of sulfur, and bonds adjacent polymeric chains.

This type of mechanism prevents polymeric chains from sliding on others that are flowing. Thus the larger the density of bonds among chains, the larger will be the material's Young's modulus.

The rotation of elastomer bonds must be rather free to ease the response of polymeric chains to stresses that cause too much strain.

If the tension of pieces of chains between bond-forming branches rises abruptly, the chains stretch and align in such a way that the secondary bonds induce crystallization. This makes the sample's Young's modulus rise considerably and creates the stiffness of elastomers.

The elastic behavior of elastomers is not perfect; therefore, there is always a small fraction

---

[4]Required tension to maintain a unit's deformation rate.

of bonds that are restructured during the application of a stress, causing hysteresis in the stress-strain curve. A stretched sample is much more ordered than if the stress had not been applied. When such stress is released there must be a disordering mechanism that leads to a rise in the sample's temperature. This rise in temperature corresponds to the energy dissipated in the stress-strain hysteresis loop.

Elastomers, like polymer networks, cannot be recycled. The gummy thermoplastic tends to flow. It is technologically interesting to find thermoplastic elastomers that can be recycled and do not flow. This can be achieved using copolymers; for example, styrene and an elastomer can coil into a central block with polystyrene chains. This triblock configuration gives rise to a phase separation when the melted material cools. Then the polystyrene agglomerates into glassy domains in an elastomer matrix. Yet each polystyrene domain can tangle several hundred polydiene chains in a way that creates a pseudonetwork. When the sample is reheated the polystyrene tangles melt and the material flows. It is possible that other elastomers with polyisoprene help to form the matrix.

Electrical Properties

Electrical Conductivity

The highly localized electrons of covalent bonds enable polymeric materials to be excellent insulators; that is, they are materials that have appreciable forbidden bands (see chapter 2), and one of their initial applications was as an insulator and dielectric; in particular, in the wrapping of copper, whose use to conduct electricity is widespread, where plastic materials substitute for previous materials. This helps in the use of coaxial wires, which are essential in the guided transmission of radio-frequency signals. This application also permits us to make boards for printed circuits, an essential component during the first stages of solid-state electronics. In addition, by combining their dielectric, thermal, and mechanical properties, they are used in the packing of optic fibers, in the low-price packaging of semiconductors and integrated circuits, and in all types of consumer electronics.

Because many polymers have low resistance to diffusion and can absorb water (hydrophilic polymers) or other fluids that can be electrolyzed, ionic conduction can be increased. Thus, application as an insulator cannot be generalized to all areas. However, ceramics are still the best insulators (although their brittleness is their biggest problem) either in situations where there are very high electric fields or at places where the temperature can be high or there are very strict technological specifications.

As we have mentioned, the behavior of a polymeric material as an insulator is influenced by factors like temperature, pollution, doping, defects, the history of the material, mechanical tensions, sample geometry, the electromagnetic field frequency, and other factors. According to the peculiarities of each application, polymeric materials present different possibilities.

Thus, PVC, PE, PTFE, and other plastics with low diffusivity are widely employed as insulators of copper conductors, as dielectrics in capacitors (table 10.3), in coaxial wires, and in the packaging of semiconductors. They are used at low breakdown voltages and at lower temperatures than are ceramics and glass. Because they are often used as dielectrics, during manufacturing any impurities that favor ionic conduction must be avoided.

Polymers whose structures have combined π electrons have special electrical properties, such as low ionization potential, low-energy optical transitions, and high electronic affinity. This makes them materials in which oxidation or reduction can take place more easily

Table 10.3 Relative permittivity (direct current) and refractive index of some common polymeric materials

Material	ϵ_r	n
Polyolefins		
Polyethylene	2.3	1.5
Polypropylene	2.2	1.49
Polyisobutylene	2.2	1.49
Vinyls		
Polyvinyl chloride	3.1	1.54
Polyvinyl fluoride	8.5	1.42
Aromatics		
Polystyrene	2.6	1.6
Polycarbonate	2.5–3.0	1.6
Nitriles		
Polyacrylonitrile	3.1	1.53
Acrylates		
Polymethyl methacrylate	2.8–3.7	1.51
Polyamides		
Polyhexamethylene (nylon 6/6)	4.1	1.50
Polyacrylamide	5.7	1.53

than in other polymers. Doping with agents that transfer a charge can convert insulating polymeric materials into conductors with a conductivity that is sometimes equivalent to that of a good metallic conductor. Doping compounds like AsF_5, I_2Li, Li, K, etc., are used and can be added by chemical or electrochemical means. Table 10.4 summarizes the conductivity values of the most common plastics and indicates their range.

Piezoelectricity and Pyroelectricity

Polymeric materials, like some ceramics (TiO_3Ba or TiO_3Sr), conserve a residual polarization without an applied electrical field, once one has been applied. Given what was said in chapter 6, this residual polarization, added to the surface charge of the material's dipoles, makes some materials piezoelectric. If an electrical field is applied in the direction of the polarization, the material's dimensions are modified in proportion to the field and the initial values are recovered when the excitation is over. Using this effect, an electrical signal transducer can be made that can produce continuous or alternating movements with high precision and speed.

Classic examples of piezoelectric polymers are polyvinylidene fluoride (PVDF), PTFE, polyvinylidene chloride, polyvinylidene cyanate, etc. To polarize them an electrical field is applied when hot, and this orients the dipoles in the field direction. When the electrical field is removed, the material conserves its polarization below a critical temperature. Owing to this property these materials are often applied in the manufacture of small microphones and loudspeakers. If a temperature higher than the critical one is reached, the system becomes disordered and this property disappears.

Table 10.4 Electrical conductivity (σ) of polymeric materials

Material	$\sigma\,(\Omega \cdot cm)^{-1}$
Dielectrics	
Polystyrene	$10^{-18} - 10^{-16}$
Polyester	
Polyimides	
Polytetrafluoroethylene	
Nylon	10^{-13}
Polymers with metallic chelates	10^{-10}
Semiconductor polymers (conjugated polymers)	
Cis-$(CH)_x$	10^{-9}
Pyropolymers	$10^{-8} - 10^{-7}$
Trans-$(CH)_x$	10^{-5}
Conductive polymers (conjugated and doped polymers)	
Polyacetylene (AsF_5, I_2Li, K)	500–2000
Polyphenylene (AsF_5, Li, K)	500
Polypyrrole	600
Polythiophene	100

Pyroelectricity depends on the material's capacity to modify this residual polarization and return to the initial conditions after receiving a thermal shock. Pyroelectric materials are normally employed as radiation sensors that are (relatively) independent of the wavelength.

Optical Properties

As good insulators, polymeric materials usually have appreciable forbidden zones that would make ideal monocrystalline phases transparent for a wide range of visible electromagnetic radiation. But the transparency can be hidden due to shape anisotropies created in the grain boundaries in semicrystalline phases or in inhomogeneities of glassy or amorphous phases. For a material to be transparent it has to be homogeneous on a length scale much larger than the wavelength of the visible electromagnetic radiation. A typical crystallite has this property. Yet because there are a great number of them in the sample, the grain boundaries cause multiple refractions and diffuse the light. This turns semicrystalline polymeric materials white or translucent, differentiating them from transparent amorphous polymeric materials. We now know that when, for example, LDPE is melted its transparency rises.

The refractive index n of the most common polymeric materials is around 1.5 (table 10.3).

Amorphous thermoplastics have contributed much in applications where a material's transparency is essential; the low cost allows for materials nearly as transparent as common glass but much stronger. The most significant example of a polymeric material used as a substitute for glass[5] is polymethyl methacrylate (PMMA or acrylic), although it has less hardness. The strength can be increased by adding compatible elastomer layers (which are quite resilient) with the same refractive index, but this reduces the material's hardness.

---

[5] Many times it is called *organic glass*.

Polystyrene is similar but less strong than PMMA. In applications where impact strength is more important than transparency—yet we want to maintain a degree of it—polycarbonate (PC) has been substituted for PMMA. There is, however, the disadvantage of being slightly less transparent, and turning yellow and brittle when exposed to ultraviolet light unless a radiation filter is used. Polyvinyl chloride, with a higher degree of crystallization, is hardly transparent and is used in packing and in containers due to its low cost. Few heat-hardened polyesters and epoxy resins are transparent.

Chemical Properties

One of the properties investigated in polymeric materials is their resistance when chemical agents are in contact with them.

Polymeric materials usually resist the attack of chemical agents such as acids and bases in aqueous media, but they are sensitive to organic solvents, which can either dissolve the samples completely or produce a plasticizing or swelling of the sample. This general view must be stated more specifically by noting that the solvent's molecules must be absorbed and diffused in the sample as a result of their attraction by parts of polymeric chains being the same as or stronger than the attractions of the chains among themselves. If diffusion does not occur, the attack will take place on the sample surface. Amorphous areas will allow more diffusion than crystallines. Thus, crystallization gives resistance to chemical agents.

A chemically aggressive environmental stress can increase the sensitiveness of polymers to fissures and fractures. Thus, solvents act preferably at the ends of microfractures and in empty spaces, where diffusion is likely to take place, causing a reduction of the ductility and allowing the microfractures to grow easily. This phenomenon is known as fracture by corrosive or environmental stress.

Damage to a sample's properties is called degradation. This degradation happens either in the manufacturing and processing of a sample, or in the environment. The maximum degradation potential is normally produced in processing because that is when the material must bear the highest temperatures and stresses. This is a technical problem, since the breakage of chains does not necessarily modify the material's structure, but does change its mean molecular mass, which reduces the viscosity in the melt, and other mechanical properties such as toughness and impact strength. This is important when recycling thermoplastics; if the proportion of irrecoverable material is not controlled, the properties of a recycled material get worse during each recycling. When a material's degradation happens over a long time it is called aging.

The oxidation of polymeric materials is complicated and normally shows when bonds appear among the chains and form networks in a sample. This implies that the sample becomes more brittle. The oxidation is often catalyzed by sunlight (usually ultraviolet); hence, to avoid oxidation either light filtration or antioxidizing pigments must be used.

10.7 DESIGN POLYMERS

We briefly describe high-performance polymers, polymeric liquid crystals, functional polymers, composite polymers, adhesives, and ionic interchange resins.

High-Performance Polymers

Polymeric materials are considered to be high performance because they have high elastic moduli and good temperature resistance. Two obvious examples of such polymers are polyimides and polyether ether ketone (PEEK); the latter has a 330°C melting point. Both serve as matrices in composite materials with carbon fibers.

Polymeric Liquid Crystals

Polymeric liquid crystals are liquid crystals (see section 12.2) built of polymeric chains. One of the most interesting lyotropic liquid crystals is the aromatic polyamide polypara-phenyilene tert-eftalamide or Kevlar fiber (PPTA), which is a solution of the polymer in concentrated sulfuric acid. This fiber decomposes at a lower temperature than that of the transition to an isotropic fluid. Another polymer with this kind of property is polyparapheny-lene benzobiaxozol (PBO). Thermotropic liquid crystals have a smaller viscosity than other melted amorphous polymeric materials, but the remaining crystallization permits closer packing.

Functional Polymers

The most important functional polymeric materials are ones that are electroactive, photoac-tive, intrinsically conducting, semiconducting (modified polyacetylene), piezoelectric, or pyroelectric. All are essential in modern engineering because they solve challenging prob-lems in applications. Their costs are low, considering that they have less weight and more strength than alternative materials. The applications are varied: display devices, optical information storing devices, xerography, laser printers, charge accumulators, transducers. All these materials have been mentioned in our treatment of the corresponding properties.

Composite Polymeric Materials

Composite or multiphase polymeric materials are a mixture of two polymeric materials or a mixture of a polymeric material and another kind of material. Polymeric materials can themselves be considered polymeric alloys because they are polydisperse. In metallic materials it is common to alloy metals to create properties intermediate between those of the components, or at least different ones. True polymeric alloys are rare among polymeric amorphous materials and are unknown among crystallines. Homogeneous alloys, like random copolymers, have only one glassy transition temperature. Although the mixing is effective, the homogeneous alloy is in a metastable state, the phases separating easily because the mutual solubility among different polymeric chains is small. Heterogeneous alloys formed of various phases have several glassy transition temperatures and can be used to make a composite material with new properties.

The outstanding examples of such composite polymers are high-impact polystyrene (HIPS) and acrylonitrile-butadiene-styrene (ABS). In HIPS a nonreticulated elastomer, usually butadiene, is dissolved in styrene and the mixture is polymerized. It must be stirred vigorously to isolate the particles—a fraction of a micrometer in size—of the reticulated elastomer spread throughout the matrix of PS, to avoid the opposite situation. In ABS the two phases are monophasic copolymers, styrene acrylonitrile (SAN) and a slightly retic-ulated butadiene elastomer. Here also the elastomer forms a fine dispersion of particles bonded into the copolymer matrix. Other, more complicated, composite polymers can be PVC, with ABS particles included or other thermoplastics toughened by an elastomer. In all cases the objective is to enhance the strength of the pure initial materials. Cellular polymeric materials are composite polymeric materials in which one of the phases is air. These types of materials are porous and are widely used in packaging, and as thermal, sound, and mechanical insulation. The main feature of these materials, shared by all porous materials, is the equal dependence of their physicochemical properties on their metric, and consequently topological, structure and on the topological nature of the *pores* found in the intrinsic polymeric material.

The first cellular polymeric material was natural rubber, which, by mechanical whisking, enfolds air. Nowadays the manufacturing methods consist of introducing inside the polymeric material a blowing agent that generates gas, for instance, when heating a material. In an open cellular structure the gas is quickly replaced by air. The most common materials are based on PS and polyurethanes. We can obtain materials from heat's insulating properties with closed cellular structures that are good thermal insulators, and with open cellular structures that are highly flexible. One useful characteristic of cellular polymeric materials is that they have a low density, which allows them to be used as a filling structure. In simple manufacturing this consists of filling a mold with a polymeric material to which a sponging agent is added. The material solidifies fast close to the mold, giving a nonporous solid polymeric layer, which keeps the temperature inside high longer and lets the sponging agent act, whereby a polymeric foam forms inside. The result combines a tough but light structure.

Adhesives

Adhesives are usually sold in liquid form. They must display a good wetting of the surfaces to be adhered and low viscosity to fill the surfaces enough so they adhere. Most adhesives are polymeric solutions that set by evaporation or diffusion of a volatile solvent. Normally, a network of bonds appears. In adhesive tape, highly plasticized gummy amorphous polymeric materials are used. They remain in a cohesive liquid state. Another type of adhesive is employed when hot; for example, ethylene–vinyl acetate is used in packing and book binding. This type melts and resolidifies during cooling.

Adhesives with network structures do not gain and lose their properties by the action of heat or solvents. This is why, when it is necessary to have adherence, even at high temperatures or in chemically aggressive atmospheres, polymeric materials, thermosets, are employed, like epoxy resins. Elastomeric material is added when it is necessary to increase strength, which is as important as the adhesive-substrate sticking strength.

Adhesives of cyanocrylate type consist of a solution of monomers stabilized by traces of an acid. When these adhesives find a slightly basic surface, the stabilizer is neutralized and the monomers quickly polymerize, enabling the substance to have strong adherence. The drawback is that there seem to be few network structures.

Finally, we have the silicones, which are elastomers that create bonds in reaction with an organometallic of tin (usually tin dibutyl diacetate). This starts to hydrolyze, evaporating acetic acid and creating the bond's precursor.

Ion Exchange Resins

Some spatial phenolic resins, discovered by Adams and Holmes in 1935, are used as cation exchange resins, acting similarly to inorganic zeolites and organic humus. One example is the resin formed when condensing phenolsulfonic acid with formaldehyde. It is often represented as an irregular mass that has many ionized sulfonic groups.

To make ion exchange resins, usually reticulated resins with a small proportion of divinyl benzene are used. Cation exchange resins are regenerated by washing with an acid solution (normally hydrochloric acid), and anion exchange resins are regenerated by washing with alkali (normally sodium hydroxide).

Although the resin is electrically neutral, when it is washed with hydrochloric acid (5%) dissolved in water, the H^+ ions move cations occupying the sulfonic groups. If in these conditions hard water is made to flow through the resin, the substitution of the H^+ ions by Ca^{++} ions demineralizes the water. The resin is regenerated when washed with an acid solution.

Some polymeric resins are obtained by condensation of urea or aromatic aminoplastics with formaldehyde. They are used as ion exchange resins (in this case, anionic) to demineralize solutions. Their macromolecules contain aminoaliphatic or aromatic groups of the type $-NH_2$, $-NH(CH_3)$, or $-N(CH_3)_2$. In general, they are represented as R-NH_2, and they are also generated by hydrochloric acid; hence, the generation of the anionic interchange resin is due to the following reaction:

$$R - NH_2 + ClH = R - NH_3(+)Cl(-),$$

the regeneration reaction being

$$R - NH_3(+)Cl(-) + HONa = R - NH_3(+)HO(-) + ClNa$$

and interchange reaction

$$R - NH_3HO(-) + anion(-) = R - NH_3(+)anion(-) + HO(-).$$

By successively passing through an acid and a basic exchange resin, a solution is demineralized. Besides their task of demineralizing water, exchange resins recover noble metals or valuable natural products from diluted solutions, and eliminate residual radioactivity.

PROBLEMS

10.1. Why does rubber crystallize when being stretched?

10.2. A polyethylene (PE) of high molecular weight has an average molecular weight of $450\ kg \cdot mol^{-1}$. What is the mean degree of polymerization?

10.3. How much sulfur must be added to 10 g of polybutadiene elastomer to cross-link 3% of the mers, assuming that all the sulfur is used in the cross-links and that only one atom is in each bond?

10.4. A polymeric material has a relaxation time of 70 days at 27°C when 7.5 MPa tension is applied. How many days are necessary for the tension to decrease to 6.2 MPa?

10.5. A polymeric material has a relaxation time of 110 days at 27°C when 5.0 MPa tension is applied. In how many days will the tension fall to 4.0 MPa?

10.6. In the previous question, what is the relaxation time at 45°C if the activation energy is $21\ kJ \cdot mol^{-1}$?

10.7. Which properties are of industrial interest in polycarbonates?

10.8. Why do thermosets usually have great strength and small ductility?

10.9. Explain the differences in mechanical properties between glassy thermoplastics and common glass.

10.10. The glassy transition temperature of polyethylene is clearly below room temperature, but the Young's modulus of high-density polyethylene is similar to that of a glassy polymeric material. Explain.

Chapter Eleven

Surface Science

Surfaces, in the widest sense, play a key role in materials science. As examples, corrosion, friction forces in solids, and catalysis of most chemical reactions happen on surfaces. Given the dimensionality of surfaces (two dimensional) and the differences between the behavior of solids in the bulk and on surfaces, one must study surfaces separately. The boom in this work led to the birth of surface science.

11.1 ELECTRONIC STRUCTURE OF SURFACES

Physicochemical processes on surfaces depend on the electronic structure and on the structure itself of the surfaces; that is, the disposition of atoms, ions, and molecules along with the surface imperfections, which can be either crystal defects or adsorbed impurities, making a "dirty surface". Before exploring this kind of real surface we should understand first perfect crystalline surfaces, without defects, without impurities, and clean. A stable surface configuration depends on the electronic structure; in metallic and semiconductor surfaces the atoms are displaced from equilibrium positions inside the material to occupy new positions of less energy. This phenomenon is called *surface reconstruction*.

Surface characterization techniques have improved enormously, and now there are more than fifty methods. Among these, several are nondestructive methods that keep the surfaces intact after measurement.

The main experimental information source for surface electronic structure is *electronic photoemission*, where the energy distribution depends on the angle of the electrons emitted by photonic excitation. Of course, the surface contribution of this photoemission should be extracted from the bulk contribution. The relation between atomic and electronic surface structures allows one to distinguish different surface geometries. Other methods for determining the surface structure are low-energy electron diffraction (LEED) and extended X-ray absorption fine structure spectroscopy (EXAFS).

Surface atoms are intermediate between those in the bulk and isolated atoms. This can be seen in the local density of states, which on the surface are composed of the exponentially decaying wave functions defined in the bulk and extra solutions for the Schrödinger equation, called the surface states. These surface states have wave functions that decay exponentially inside the material and thus are confined on the surface. This exponential decay allows one to define the material's surface more rigorously as the region comprised of the first few atomic layers (approximately three) that divide the material from another material or from the vacuum. These surface states emerge from the Bloch theorem, as with bulk materials, but they have a complex wave vector (at least in the direction perpendicular to the surface) that creates the probability of finding an electron in this state which decays exponentially when we move away from the surface. These surface states are highly dependent on surface geometry, and in semiconductors they have the characteristic of *free bonds* (chemical bonds that were broken when the surface was made). The interpretation of the electronic structure as local chemical bonds is fine in semiconductors and dielectrics, but not in conductors, where the Fermi gas removes the contribution of the orbitals of the valence electrons. Changes in the local density of states mean that, in transition metals, distinct magnetic behavior occurs on the surface. The number of surface states compared to the total number of states is of the order of the ratio of surface atoms to bulk atoms; that is, of the order of 10^{-8}. Hence, states like these are important only near the surface of bulk materials, or in quasi-two-dimensional materials, or in layer structured materials. Surfaces can have collective electronic oscillations (surface plasmons) and other collective phenomena.

It is essential to consider that the potential perceived by an electron (without taking collective phenomena into account) does not have the symmetry of a three-dimensional Bravais lattice. So we can assert that the surface crystalline structure is biperiodic; that is, the surface is not mathematically surface periodic in two dimensions. But the periodicity of the Bravais lattice bulk is lost in the perpendicular direction at the surface and may have, in the directions tangential to the surface periodicities different from those of the bulk. The bulk material is called the *substrate*. The surface may be formed of different materials or may be the edge of the substrate. The bidimensional lattice is formed by the periodic repetition of the unit cell and is called the *mesh*. If the surface is composed of a material different from the substrate, then the surface is made up of adsorbed *adatoms*. Layers of adatoms are called *adlayers*. Adatoms can be adsorbed on the surface through either weak or strong bonds, called physical or chemical, respectively. Surface structure characterization techniques range from low-energy electron diffraction to reflection high-energy electron diffraction and electronic photoexcitation, which is a special case of electronic photoemission. One of the greatest advances in surface structure detection is the scanning tunneling microscope (STM), where the electronic tunneling effect between the material's surface and the microscope's metallic tip (with a diameter of about 10 Å), placed at a distance of approximately 6 Å from the surface, allows one to map the surface's electronic structure when the whole surface is examined. The final map is related to the local density of states at the Fermi energy. In the last years of the twentieth century new techniques such as atomic force microscopy, among others, were developed and give similar information about the surface.

Related to the surface structure and the surface electronic structure is an easily measured quantity called the *work function* W, which is the energy necessary to extract an electron from a material. In general, the work function reflects changes in the surface charge distribution (figures 11.1 and 11.2) and represents the difference between the vacuum level and the Fermi level. The energy of the vacuum level is the energy of an electron at rest far away from the material (around 100 Å). The vacuum energy is such that the electrostatic force caused by the polarization induced by an extracted electron is negligible, and the energy of the Fermi level is contained in the material's electronic electrochemical potential. This is true at $T = 0$ K, and it is approximate for positive temperatures.

The various work functions of different materials allow one to observe *contact potentials* in metallic materials, which produce effects similar to the ones noted in chapter 4. If we put two conductors (with different work functions) in contact, the Fermi energies of the electrons in their respective surfaces will be different, given their different contact potentials. Hence the electrons will flow to the energetically favored region, appearing as a net charge in each conductor until the induced voltage prevents more charge flow. If two conductors are joined at two surfaces and each junction is at a different temperature, there will be electrical conduction. This is a microscopic explanation of the thermoelectric effects noted in chapter 4. Also, the work function is related to the *photoelectric effect* and

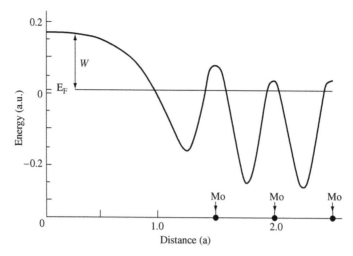

Figure 11.1 Mean potential energy in the direction perpendicular to the (001) surface of Mo. The
lattice parameter is represented by a

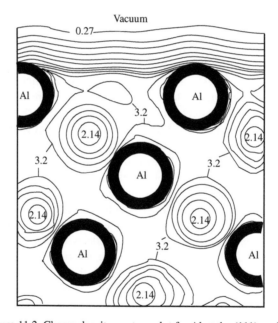

Figure 11.2 Charge density contour plot for Al at the (111) surface.

to *thermoionic emission*. The former has a threshold energy at $T = 0$ K equal to the work
function, and if the incident photon energy is $h\nu$, the kinetic energy of the emitted electron
is $h\nu - W$. The latter is the emission of electrons because the electron gas in a conductor at
a finite temperature is at a chemical potential different from that of the vacuum electron gas
(which is not very dense). In this instance, the emission of electrons balances the chemical
potentials. In other ways, thermal fluctuations of electron energy in a material may be such
that some of the electrons overcome the energy barrier of the work function and leave the
solid.

Electronic photoemission techniques, such as, for instance, the photoelectric effect mentioned above, are important for observing the electronic structure of atoms, molecules, and condensed matter (materials). When electronic photoemission is employed for the observation of this structure it is called *photoelectronic spectroscopy*, which can especially garner information about the band structure, the holes and electrons, the mean lifetimes, the energy levels of atoms, molecules, or ions adsorbed on the material's surfaces, etc. The fast growth of this science comes from the ability to clean surfaces, to keep them clean during their analysis, and to expose them to known adsorbates in a controlled way. For this it is necessary to have an ultrahigh vacuum (UHV) (10^{-10} torr). Gas exposure is measured in langmuirs ($1 \text{ L} = 10^{-6}$ torr $\cdot$ s); 1 L is the exposure that covers a surface of 1 cm^2 with a gas monolayer, assuming that all molecules impinging on the surface are adsorbed. Photoemission is quite complex. As a first approximation for low energies the model of three independent stages is valid: excitation of an electron by a photon, transport of an electron to the surface, and escape of an electron from the surface. As is clear, the effect of the surface is important and therefore the surface be studied with this kind of spectroscopy.

11.2 NANOCRYSTALLINE SURFACE FORMATION

A clean surface (i.e., without adsorbed impurities) can be either reconstructed or unreconstructed. In unreconstructed surfaces the atomic ordering agrees with that of the substrate, only changing the spacing between layers, in what is called a *relaxation multilayer* at the surface. This relaxation multilayer dominates especially in the spacing between the first and second layers; it is smaller in the following spacings. To understand this phenomenon the surface should be taken as an intermediate state between a diatomic molecule and the substrate, and where the interatomic distance is less in the former than in the latter.

In unreconstructed surfaces, therefore, the substrate structure is kept as its projection on the surface. In reconstructed surfaces the atomic position relaxations create a qualitatively different structure at the surface, with a new primitive unit cell.

Sometimes in metallic materials and often in nonmetallic ones, surfaces are reconstructed due to a restructured covalent or ionic bond at the surface, which is free at the time the surface is created. This bond will have a greater associated energy; diminishing its energy leads to atomic displacements that can reach up to 0.5 Å. In this context, external atom adsorption is favored, which also can lead to reconstructed surfaces.

The phenomenon of surface reconstruction is quite significant from a technological viewpoint because nanocrystalline growth techniques such as molecular-beam epitaxy (MBE) and chemical-vapor deposition (CVD), which we referred to in chapter 4, depend directly on the energy ratio between a nonreconstructed and a reconstructed surface. Therefore, if we have a nonreconstructed clean surface we can add an adlayer that will or will not reconstruct the surface, and has or has not the same symmetry as the surface. This kind of process is extremely important because the surface structure, or the adlayer, can change the material's mechanical, thermal, electrical, magnetic, or optical properties and improve its usefulness in applications. Hence, surface reconstruction is worthwhile for phenomena like corrosion, catalysis, and semiconductor junctions, where the surface or interface contribution to the phenomenon is major.

Metallic materials, when grown slowly, exhibit *epitaxial growth*, that is, ordered. This ordered growth has essentially three different growth modes. The first is called the Franck–Van der Merwe mechanism and consists of the consecutive deposition of different adlayers on the substrate (figure 11.3). The second is called the Stranski-Krastanov mechanism and consists of the growth of three-dimensional islands on a first adlayer (figure 11.3). The third is called the Volmer-Weber mechanism, where three-dimensional islands appear from the very beginning.

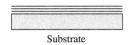

Substrate

a) Mechanism of Franck-Van der Merwe

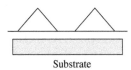

Substrate

b) Mechanism of Stranski-Krastanov

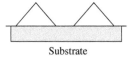

Substrate

c) Mechanism of Volmer-Weber

Figure 11.3 Mechanisms of adlayer growth. (a) Franck–Van der Merwe, (b) Stranski-Krastanov, and (c) Volmer-Weber.

The Stranski-Krastanov and Volmer-Weber mechanisms add difficulty to the perfect growth of surfaces and solids, but they allow creation of quantum dots, which are isolated three-dimensional islands. This isolation permits one to have many independent nanosystems with energy bands for which the quasicontinuum approximation is not valid, and also allows narrow energy transitions, which are useful, for example, in lasers and isolated photon detection.

PROBLEMS

11.1. Why can a clean surface be reconstructed or not reconstructed?

11.2. If a surface creates surface states in the band structure, discuss the energy band structure created by the grain boundaries in a polycrystalline material.

11.3. A porous material is generally considered a material with a huge external effective surface. What does this surface look like? Discuss its energy band structure.

11.4. Explain the basis of a STM.

11.5. What is the minimum energy that a photon needs to produce a photoelectric effect in a material?

11.6. What information can be obtained from photoelectronic spectroscopy? Discuss how all this information could be filtered.

11.7. The atomic (or molecular) flux $j = N_A P / \sqrt{2\pi M N_A K_B T}$ (in the cgs system) on a surface is proportional to the pressure and inversely proportional to the square root of the molar mass multiplied by the absolute temperature. Consider that each atom (or molecule) which impinges on the surface is adsorbed (without dissociating if it is a molecule). What is the proportionality constant at a pressure of 3×10^{-6} torr of molecules of CO at 300 K, if the molecular flux is of the order of 10^{15} in corresponding units?

11.8. Copper (100) planes adsorb 1% of the oxygen molecules that impinge on them at 300 K. Calculate the pressure needed to keep clean a surface of 1 cm^2 of copper for 1 s, 1 hour, and 8 hours.

11.9. What do all the techniques of nanocrystalline growth have in common?

11.10. Discuss what type of material would choose each epitaxial growth mode (figure 11.3).

Chapter Twelve

New Materials

Even term "new materials" is ambiguous and time dependent, so generally we use it to name materials that cannot be included in the classification of page 4. This means that fullerenes, liquid crystals, and most biocompatible materials can be called new materials. Also, other materials can be classically classified, but they have new properties and therefore can be considered new materials. Examples are new ceramics, new metals, new composite materials, and so on. In this chapter we examine only fullerenes, liquid crystals, and biocompatible materials, giving each a brief introduction.

12.1 FULLERENES

Fullerenes are allotropic carbon forms different from diamond and graphite. They are uniquely stable and finite forms. They were found in 1990 by Krätschmer and Fostiropoulos, having been predicted in 1966 by Jones and discovered, without confirmation, in 1985 by Smalley, Curl, and Kroto. Fullerenes are molecular solids composed of pseudospherical molecules containing from 32 to 960 carbon atoms. These forms increase stability because they have few free bonds. The best known, common, and stable of the fullerene-type molecules is *buckminsterfullerene* (C_{60}) (figure 12.1), which has diameter of around 1 nm.

After the discovery of buckminsterfullerene, buckybabies (C_{32}, C_{44}, C_{50}, and C_{58}) were proposed, as well as the C_{70} fullerene (also obtainable in appreciable amounts) and the gigantic fullerenes (C_{240}, C_{540}, and C_{960}). Each fullerene is a closed hexagonal lattice, with enough pentagons to permit closure of the molecule. There are 12 pentagons independent of the particular fullerene; this was calculated geometrically by the Swiss mathematician Leonhard Euler. Buckminsterfullerene has exactly 20 hexagons and C_{70} has 25.

There are also synthesized compounds that are partly fullerenes. For example, $C_{60}(OsO_4)$ (4-tert-butylpyridine)$_2$, K_3C_{60}, $C_{60}F_{60}$, and so on, have intriguing properties. Small atoms can also be included inside the fullerene molecules.

From the standpoint of materials science, the most important thing is not these molecules' structure but the structure of the macroscopic materials made with them. Thus, buckminsterfullerene crystallizes as a fcc lattice and has been predicted to be an intrinsic semiconductor with a direct band gap. At lattice sites, the fullerene molecules rotate randomly at such a high frequency that they disorder the buckminsterfullerene, so the material behaves as if it has partial order: It has global positional order, but neither orientational nor angular velocity order at each lattice site.

K_3C_{60}, a buckminsterfullerene potassium salt, displays metallic properties and has a fcc structure with a physical base formed by an atom of K and a buckminsterfullerene molecule. This material behaves as a superconductor for temperatures lower than 18 K. If potassium is substituted by rubidium or by rubidium and thalium, the critical temperature increases to 43 K.

Figure 12.1 Buckminsterfullerene molecule.

Buckminsterfullerene can be doped, for example, by putting impurities inside the molecule. This modifies its properties significantly.

Molecules of $C_{60}F_{60}$ have a stable and closed chemical structure, behaving like teflon spheres, and this material could be of technological importance as a good lubricant.

The synthesis of fullerenes led to research in creating more complex shapes like the nanotubes C_{74}, C_{76}, C_{84}, and others, all of which have promising technological applications.

12.2 LIQUID CRYSTALS

When heated, some crystalline solids go through intermediate phases before acquiring the typical characteristics of isotropic liquids. These phases are called mesophases, and the materials showing them are generically called *liquid crystals*. Mesophases have partial order and therefore behave like crystalline solids and flow as common liquids, in phases that are not solid. This is how they got their name. Liquid crystals were discovered by Reinitzer (1888).

Liquid crystals can be placed into two groups, depending on how the mesophases are obtained. *Thermotropic liquid crystals* are those that form mesophases by changes in temperature; and *lyotropic liquid crystals* are those whose multicomponent liquid crystals attain mesophases when their concentration is changed.

Mesophases with partial order are due to the large anisotropy of their components. Hence, it is possible to have liquid crystals with molecules that have barlike shapes, like aromatic organic molecules. There are also mesophases in compounds with disklike molecular shapes and in some polymeric materials.

The most important phases of a liquid crystal when its temperature and its concentration, or both, change are solid, isotropic liquid, and gaseous, and the mesophases. Solids are crystalline solid phases and possess complete order at low temperatures; gases correspond to high temperatures and are completely disordered. Mesophases can be classified as follows (figure 12.2)

1. *Nematic.* In this mesophase the liquid crystal molecules display nonpolar orientational order but not positional order. The orientational order defines a director vector, which is the mean molecular orientation in a domain, called a volume element or fluid particle, leading to all the anisotropy presented in this mesophase: refractive index, permittivity, and permeability anisotropies. Positional disorder means that the molecules are not at the sites of a hypothetical Bravais lattice and thus flow almost like a common liquid does, but they have anisotropic viscosity coefficients.

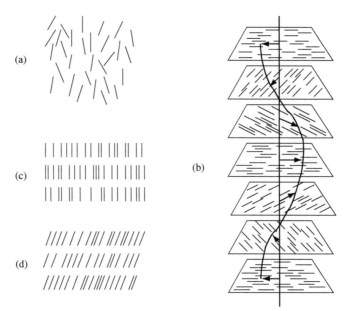

Figure 12.2 Various mesophases of liquid crystals. (a) Nematic, (b) cholesteric, (c) smectic A, and
(d) smectic C phase.

2. *Cholesteric.* This mesophase is formed by optically active molecules in a way similar
to nematic liquid crystals. The arrangement of the volume elements in a sample consists of
layers oriented in one of the planar directions formed by the layer; this direction changes
continuously and has a helical structure. The step, defined as the distance between the
plane of one orientation and the nearest plane with the same orientation, is of the order of
magnitude of the wavelength of visible electromagnetic radiation. From this structure arise
most applications of the cholesteric phase.

3. *Smectics.* Smectics are mesophases characterized by having orientational and positional order in one dimension; that is, they are nematic-like mesophases but layer structured.
If the layers have the width of a volume element and are perpendicular to other layers, then
the liquid crystal phase is smectic A. If the volume elements are tilted, the resulting phase
is smectic C. Other smectic phases have layers with widths greater than a volume element,
or have a certain order inside each layer.

4. Other mesophases like blue phases, and so on.

Typical examples of phase transitions in common liquid crystals are, for PAA (*p*-
azoxyanisole), crystal $\rightarrow$ 118.2°C $\rightarrow$ nematic $\rightarrow$ 135.3°C $\rightarrow$ isotropic liquid; for 5CB
(pentylcyanobiphenyl), crystal $\rightarrow$ 22.5°C $\rightarrow$ nematic $\rightarrow$ 35°C $\rightarrow$ isotropic liquid; for N-
(*p*-heptyloxybenzylidene)-*p-n*-pentylaniline, crystal $\rightarrow$ smectic H $\rightarrow$ smectic B $\rightarrow$ smectic
C $\rightarrow$ smectic A $\rightarrow$ nematic $\rightarrow$ isotropic liquid.

The properties of liquid crystals can be summarized as follows.

- Birefringence. Liquid crystals have a refractive index that depends on the electrical
  field's direction of electromagnetic radiation. Thus, it will be different for volume
  elements corresponding to the axis of the director vector and those perpendicular to
  this axis.

- Anisotropy in permittivity, permeability, an electrical conductivity.

- Anisotropy in thermal conductivity.

- Anisotropy in viscosity and elasticity. Liquid crystals exhibit viscoelastic behavior when a material is strained, which indicates the established order (for instance, orientational).

- Surface alignment. The orientation of volume elements is strongly determined by the boundary conditions.

The applications of liquid crystals are derived from the properties of these materials. The evolution of applications has been quite important recently; as an example, there are several ways to make liquid crystal displays based on different properties. For instance, an electrical field can rotate a volume element, which in turn can change the light polarization depending on its orientation. Then a matrix of liquid crystal is put between two crossed polarizers. There we can excite the crystal, pixel by pixel, and through a color filter obtain the colors red, green, and blue. In this way we can selectively pass different colors and intensities from the light source to the front of the screen. Also, we can change the step of a cholesteric crystal with an electrical field in such a way that the color of the reflected light depends on the step. In a similar way we can alter the step using a temperature field, which permits the construction of thermometers in which each color represents a different temperature.

12.3 BIOCOMPATIBLE MATERIALS

Biocompatible materials—sometimes called biomaterials—are ones that, even without being of biological origin, can coexist long enough to complete their functions in contact with part of a living organism (usually a human being). Biocompatible materials can carry out a range of functions from tissue and natural organ replacements to social functions (which we will not discuss), and the support of these tissues or organs in their normal activities. In this section we explore materials that require high compatibility, and we skim over the others.

If a material must be biocompatible, it is first of all necessary that it be nontoxic for the living creature that is the host. This means that toxic residue should not be released, and also that the material should not change other biological materials (or their properties) in the living organism, such as, for example, the sedimentation velocity of blood. Finally, it is generally necessary that in the short term, the living creature does not recognize the biocompatible material as a strange body. All these requirements mean that biocompatible materials do not belong to any classifications of materials cited above. Since their most important aspect is their biocompatibility, they are classified according to their compatibility and hence their specific function. That is, a material can be biocompatible when attached to skin but not when it substitutes for the heart.

In addition to this necessary biocompatibility, with regard to materials engineering it is fundamental that biomaterials should efficiently perform the main functions of the natural organs or tissues that they replace or support. In this sense, it is necessary that biomaterials have appropriate mechanical properties (resistance to impact, stiffness, elastic limit), thermal properties (thermal conductivity, thermal expansion, specific heat), and electrical properties (conductivity, permittivity, and so forth). All these properties should be valid across the entire range of environmental conditions and specific use. A key factor is material degradation. In general it is desirable that the degradation time is much longer than the living creature's mean lifetime; and when degradation could be useful, it should lead to residues that can become assimilated and expelled, or both (secreting or excreting them), like self-absorbing sutures.

The level of biocompatibility required depends on the part of the living creature with which the material interacts as well as its functions. The different requirements can be summarized as follows.

1. Materials requiring very little biocompatibility. These materials generally are in contact with the skin, so they should not contain allergens and should not give off toxic residues. Among these there are natural materials like cotton, wool, silk, and flax, or other materials like polyamides, polyesters, and others that insulate us from heat and cold. Moreover, they should be noninjurious when used in devices that improve the senses, like vision or hearing, or compensate for sensory dysfunction. Examples of this kind are external prostheses for hearing and vision. Generally these are not regarded as new materials because they can be classified in the usual groups of materials.

2. Materials requiring little biocompatibility. Generally such materials are in contact either with mucus membranes or with organs or tissues insensitive to strange bodies. The required conditions are similar to the preceding ones but at a higher level. The materials could be rubber, noble metals, ceramics, or polymeric materials whose main uses are as dental prosthesis, medical examination gloves, surgical instruments, diagnosis instruments (diaphotoscopes, for instance), contact lenses, and so on. At this level we have subdermal implants and transdermal patches also. Even though these need only a little compatibility, they should behave as high-technology membranes and have properties like other new materials.

3. Materials requiring either moderate or high biocompatibility. The two cases are differentiated depending on the gravity of incompatibility.

This kind of material was initially of biological origin, for example, transplants (heart, lung, kidney, skin, liver, bone, bone marrow, blood), where doctors use organs or tissues from living creatures. Among these transplants there are three types: self-transplants (from the same organism), homotransplants (from creatures of the same species), and heterotransplants (from organisms of other species). When the first option is possible, it is the best one. The second is better than the third because the functions adjust better and the tissues are slightly more compatible. Some transplants (e.g., blood transfusion) have only minimal requirements: the same blood group and Rh factor give nearly complete compatibility. Other transplants, however, make it necessary to avoid rejection by administering immune suppressors.

Materials that do not have a biological origin need more complex adaptation. Specific properties of materials are required—one of the most critical is durability. Also, biocompatibility can be attained in two different ways. The first consists of using inert materials. The second relies on interactive materials that create chemical bonds with the surrounding tissue or are covered spontaneously by biological tissues. The resulting interface is called a pseudointima and is a biologically stable approximation to a natural covering. Here we examine only the following four cases.

The cardiovascular system. Here, in addition to the need for high biocompatibility, materials that replace parts must have the elasticity and fatigue resistance required to beat 40 million times a year, thereby avoiding the need to operating again to replace any part soon. These materials should not damage blood cells or form clots by having proteins deposited on them. Solutions, at least for these problems, are modified silicones, even though the resistance and durability are limited, polyurethanes like Biomer® with a base of polyether, or pyrolytic carbon in parts where flexibility is not required. Normally, when a human heart is replaced, these materials and others are combined to achieve functionality without losing biocompatibility. Other materials to use are textured polyurethanes, to which proteins, platelets, and fibrin adhere. Fibroblasts then remain at the interface, synthesizing collagen at a stable and smooth surface. Polytetrafluoroethylene has also been used.

Artificial blood is one of the most difficult materials to find; it must combine biological compatibility with the ability to perform the key functions of natural blood. One of these functions—carrying oxygen efficiently—can be carried out by perfluorocarbons.

Artificial skin. Natural skin comprises several layers with distinct functions, so it is quite complicated to come up with a material to replace it. Among the necessities are transpiration and protection against infectious diseases. Hence, we need a composite material. The difficulty can be overcome with a transitory artificial skin that has two objectives: to stimulate a new dermis that can grow, and to protect us against infections. Once the new dermis has grown, the artificial skin deteriorates and loses its usefulness. Afterward, the artificial skin is removed and pieces of epidermis (from the same patient) are implanted such that in a few days the new epidermis will grow. It is also possible to grow semiartificial *skins* in laboratory-controlled cultures and implant them on patients.

Artificial bone. The materials used in bone implants should have a high stiffness and load resistance. Bone remineralization is very important, and implants favor it; the implant dissolves and leaves recovery up to the natural bone. Initially, these prostheses were metallic (titanium, chromium, and cobalt alloys). The main problem with this kind of prosthesis is that it supports the load passively. But the natural bone regenerates itself and grows because of the load. Thus, near the prosthesis the lower stresses exerted on the surrounding bone can reduce it. A partial solution is to cover the metallic part of the prosthesis with calcium phosphate, glass ceramics, or biocompatible glasses (where a portion of the silicon is replaced by calcium, phosphorus, and sodium), which can join tightly with the natural bone, thereby favoring deposits of natural bone components.

Collagen. A new family of materials of organic origin (basically proteins) are used in food packaging, usually meat packaging. They are biodegradable and hence are tolerated well by the human digestive system. The main disadvantage is the difficulty of sealing it—a problem not yet fixed.

Appendix A

Physical Constants

- Absolute value of the electron electric charge $e = 1.602 \times 10^{-19}$ C.

- Avogadro's number $N_A = 6.023 \times 10^{23}$ mol^{-1}.

- Bohr magneton $\mu_B = e\hbar/2m_e$.

- Boltzmann constant $K_B = 1.380 \times 10^{-23}$ J $\cdot$ K^{-1}.

- Elemental magnetic flux $\Phi_0 = h/2e$.

- Gas constant $R = N_A K_B$.

- Mass of the electron $m_e = 9.109 \times 10^{-31}$ kg.

- Permeability of vacuum $\mu_0 \equiv 4\pi \times 10^{-7}$ N $\cdot$ A^{-2}.

- Planck constant $h = 6.626 \times 10^{-34}$ J $\cdot$ s.

- Reduced Planck constant[1] $\hbar = h/2\pi$.

- Speed of light in vacuum $c \equiv (\epsilon_0\mu_0)^{-1/2} \equiv 2.99792458 \times 10^8$ m $\cdot$ s^{-1}.

---

[1] It is read "*h bar*."

Bibliography

[1] **Anderson, J. C., Leaver, K. D., Rawlings, R. D., and Alexander, J. M.** 1990. *Materials Science.* 4th ed., Chapman & Hall, London.

[2] **Ashcroft, N. W., and Mermin, N. D.** 1976. *Solid State Physics.* Saunders College Publishing, Philadelphia.

[3] **Piel, J. (ed.),** 1986. Thematic issue on new materials, Scientific American 255 (4).

[4] **Callister, W. D.** 2002. *Materials Science and Engineering: An Introduction.* 6th ed., Wiley, New York.

[5] **Dorlot, J., Baïlon, J., and Masounave, J.** 1991. *Des Materiaux.* 2nd ed., Editions de l'École Polytechnique de Montréal, Montréal.

[6] **Kittel, C.** 1996. *Introduction to Solid State Physics.* 7th ed., Wiley, New York.

[7] **Pávlov, P. V., and Jojlov, A. F.** 1992. *Física del Estado Sólido.* Editions Mir, Moscow.

[8] **Perry, R. H., Green, D. W., and Maloney, J. O.** 1997. *Perry's Chemicals Engineers' Handbook.* 7th ed., McGraw-Hill, New York.

Specific Bibliography

[9] **Harrison, W. A.** 1980. *Solid State Theory.* Dover, New York.

[10] **Warren, B. E.** 1990. *X Ray Diffraction.* Dover, New York.

[11] **Nabarro, F.R.N.** 1987. *Theory of Crystal Dislocations.* Dover, New York.

[12] **Adler, R. B., Smith, A. C., and Longini, R. L.** 1964. *Introduction to Semiconductor Physics.* Wiley, New York.

[13] **de Groot, S. R., and Mazur, P.** 1984. *Nonequilibrium Thermodynamics.* Dover, New York.

[14] **Craik, D. J.** 1995. *Magnetism: Principles and Applications.* Wiley, New York.

[15] **Jiles, D.** 1998. *Magnetism and Magnetic Materials.* 2nd ed. Chapman & Hall, London.

[16] **Giaver, I., and Megerle, E.** 1961. Physical Review 122, 1101–1111.

[17] **Rose-Innes, A. C., and Rhoderick, E. H.** 1978. *Introduction to Superconductivity.* 2nd ed., Pergamon, Oxford.

[18] **Tallon, J.** 2000. Physics World 13 (11), 25–26.

[19] **Hecht, E.** 2002. *Optics.* 4th ed., Benjamin Cummings. San Francisco, CA.

[20] **Angell, C. A.** 1995. Formation of glasses from liquids and biopolymers. Science 267, 1924–1935.

[21] **Greer, A. L.** 1995. Metallic glasses. Science 267, 1947–1953.

[22] **Janot, C.** 1997. *Quasicrystals: A Primer.* Oxford University Press, Oxford.

[23] **Stillinger, F. H.** 1995. A topographic view of supercooled liquids and glass formation. Science 267, 1935–1939.

[24] **Santen, L., and Krauth, W.** 2000. Absence of thermodynamic phase transition in a model glass former. Nature 405, 550–551.

[25] **Saika-Voivod, I., Poole, P. H., and Sciortino, F.** 2001. Brittle-to-strong transition and polyamorphism in the energy landscape of liquid silica. Nature 412, 514–517.

[26] **Baer, E.** 1986. Advanced polymers. Scientific American 255 (4), 178–190.

[27] **Boyd, R. H., and Phillips, P. J.** 1996. *The Science of Polymer Molecules.* Cambridge University Press, Cambridge, England.

[28] **David, D. J., and Misra, A.** 2001. *Relating Materials Properties to Structure.* Technomic, Lancaster, PA.

[29] **Findley, W. N., Lai, J. S., and Onaran, K.** 1994. *Creep and Relaxation of Nonlinear Viscoelastic Materials.* Dover, New York.

[30] **Frick, B., and Richter, D.** 1995. The microscopic basis of the glass transition in polymers from neutron scattering studies. Science 267, 1939–1945.

[31] **Martínez de las Marías, P.** 1972. *Química y Física de los Altos Polímeros y Materias Plásticas.* Ediciones Alhambra, Madrid, Spain.

[32] **Richards, R. B.** 1951. Journal of Applied Chemistry 1, 370.

[33] **Somorjai, G. A.** 1994. *Introduction to Surface Chemistry and Catalysis.* Wiley-Interscience, New York.

[34] **Fuller, R. A., and Rosen, J. J.** 1986. Materials for medicine. Scientific American 255 (4), 118–125.

Index

GPSR Authorized Representative: Easy Access System Europe - Mustamäe tee
50, 10621 Tallinn, Estonia, gpsr.requests@easproject.com